Air Pollution: Measurement, Impacts and Control

Air Pollution: Measurement, Impacts and Control

Kylan Wilkins

www.callistoreference.com

Callisto Reference,
118-35 Queens Blvd., Suite 400,
Forest Hills, NY 11375, USA

Visit us on the World Wide Web at:
www.callistoreference.com

ISBN: 978-1-64116-227-2 (Hardback)

Cataloging-in-Publication Data

Air pollution : measurement, impacts and control / Kylan Wilkins.
  p. cm.
Includes bibliographical references and index.
ISBN 978-1-64116-227-2
1. Air--Pollution. 2. Air--Pollution--Measurement. 3. Pollution prevention.
4. Environmental impact analysis. 5. Environmental monitoring. I. Wilkins, Kylan.
TD890 .A37 2019
628.53--dc23

Table of Contents

Preface

When excessive quantities of pollutants or harmful substances such as particulates, gases, etc. are introduced into the Earth's atmosphere, it leads to air pollution. It is damaging to the natural and built environment and may potentially result in allergies, diseases and death in humans. Pollutants can be natural or man-made. They can be further classified into primary and secondary pollutants. Carbon dioxide is a primary pollutant that has the most significant contribution to air pollution. Others include sulphur oxides, nitrogen oxides, persistent free radicals, chlorofluorocarbons, etc. Some secondary pollutants are ground-level ozone and peroxyacetyl nitrate. Reduction of fossil fuel use, transition to renewable and clean energy, use of particulate control devices, etc. are some of the techniques for the control of air pollution. The air quality index (AQI) is a measure of how polluted the air is. This book elucidates the concepts and innovative models around prospective developments with respect to air pollution control. It aims to shed light on some of the unexplored aspects of air pollution and its impacts. It is an essential guide for both academicians and those who wish to pursue this discipline further.

To facilitate a deeper understanding of the contents of this book a short introduction of every chapter is written below:

Chapter 1- When contaminants such as harmful gases, particulates and biological molecules enter the Earth's atmosphere, it results in air pollution. The activities responsible for the release of such harmful contaminants are varied and can be both natural and anthropogenic. This is an introductory chapter, which will introduce briefly the different types of air pollution and their causes.

Chapter 2- Air pollutants are the substances that cause pollution in air. These can be gaseous, liquid droplets or solid particles. The classification and major types of air pollutants such as hydrogen sulphide, oxides of nitrogen, sulphur dioxide, carbon monoxide, etc. and their impacts have been discussed in this chapter.

Chapter 3- Indoor air quality is the quality of air within and around human habitations. It can be adversely affected by the presence of gases, microbes, etc. The diverse topics elaborated in this chapter include ventilation, indoor bioaerosols and passive smoking, which will help in developing a better perspective about indoor air quality.

Chapter 4- An important index for communicating the actual or predicted measure of air pollution is the air quality index. It is obtained by computing the air pollutant concentration over a specified averaging period. This chapter discusses in detail the different ways to measure air quality, such as air quality index, pollutant standards index, NowCast, Beta attenuation monitoring, etc.

Chapter 5- Various technologies and strategies are available for the reduction of air pollution. The aim of this chapter is to explore the various air pollution mitigation technologies, such as the use of thermal oxidizer, emissions trading, dust collection system, diesel particulate filter, etc.

Chapter 6- Scrubber systems are pollution control devices that remove gases and particulates from industrial exhaust streams. Science and technology have undergone rapid developments in the past decade, which has resulted in the discovery of significant tools and techniques such as wet scrubbers, carbon dioxide scrubbers and spray towers, which have been extensively detailed in this chapter.

I would like to share the credit of this book with my editorial team who worked tirelessly on this book. I owe the completion of this book to the never-ending support of my family, who supported me throughout the project.

Kylan Wilkins

Chapter 1

Air Pollution: An Introduction

When contaminants such as harmful gases, particulates and biological molecules enter the Earth's atmosphere, it results in air pollution. The activities responsible for the release of such harmful contaminants are varied and can be both natural and anthropogenic. This is an introductory chapter, which will introduce briefly the different types of air pollution and their causes.

Air lets our living planet breathe—it's the mixture of gases that fills the atmosphere, giving life to the plants and animals that make Earth such a vibrant place. Broadly speaking, air is almost entirely made up of two gases (78 percent nitrogen and 21 percent oxygen), with a few other gases (such as carbon dioxide and argon) present in absolutely minute quantities. We can breathe ordinary air all day long with no ill effects, so let's use that simple fact to define air pollution, something like this.

Air pollution is one such form that refers to the contamination of the air, irrespective of indoors or outside. A physical, biological or chemical alteration to the air in the atmosphere can be termed as pollution. It occurs when any harmful gases, dust, smoke enters into the atmosphere and makes it difficult for plants, animals and humans to survive as the air becomes dirty.

Air pollution can further be classified into two sections- Visible air pollution and invisible air pollution. Another way of looking at Air pollution could be any substance that holds the potential to hinder the atmosphere or the well being of the living beings surviving in it. The sustainment of all things living is due to a combination of gases that collectively form the atmosphere; the imbalance caused by the increase or decrease of the percentage of these gases can be harmful for survival.

The Ozone layer considered crucial for the existence of the ecosystems on the planet is depleting due to increased pollution. Global warming, a direct result of the increased imbalance of gases in the atmosphere has come to be known as the biggest threat and challenge that the contemporary world has to overcome in a bid for survival.

The Earth is surrounded by a blanket of air (made up of various gases) called the atmosphere. The atmosphere helps protect the Earth and allow life to exist. Without it, we would be burned by the intense heat of the sun during the day or frozen by the very low temperatures at night.

Things that pollute the air are called pollutants. Examples of pollutants include nitrogen oxides, carbon monoxides, hydrocarbons, sulphur oxides (usually from factories), sand or dust particles, and organic compounds that can evaporate and enter the atmosphere.

There are Two Types of Pollutants

Primary pollutants: Primary pollutants are those gases or particles that are pumped into the air to make it unclean. They include carbon monoxide from automobile (cars) exhausts and sulfur dioxide from the combustion of coal.

Before flue-gas desulfurization was installed, the emissions from this power plant in New Mexico contained excessive amounts of sulfur dioxide.

Secondary pollutants: When pollutants in the air mix up in a chemical reaction, they form an even more dangerous chemical. Photochemical smog is an example of this, and is a secondary pollutant.

Substances emitted into the atmosphere by human activitty include:

- Carbon dioxide (CO_2) – Because of its role as a greenhouse gas it has been described as "the leading pollutant" and "the worst climate pollution". Carbon dioxide is a natural component of the atmosphere, essential for plant life and given off by the human respiratory system. This question of terminology has practical effects, for example as determining whether the U.S. Clean Air Act is deemed to regulate CO_2 emissions. CO_2 currently forms about 410 parts per million (ppm) of earth's atmosphere, compared to about 280 ppm in pre-industrial times, and billions of metric tons of CO_2 are emitted annually by burning of fossil fuels. CO_2 increase in earth's atmosphere has been accelerating.

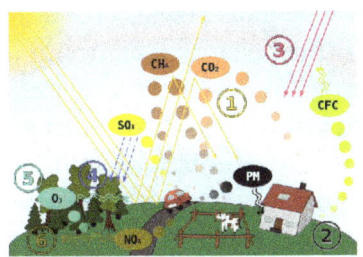

Schematic drawing, causes and effects of air pollution: (1) greenhouse effect, (2) particulate contamination, (3) increased UV radiation, (4) acid rain, (5) increased ground level ozone concentration, (6) increased levels of nitrogen oxides.

Thermal oxidizers are air pollution abatement options for hazardous air pollutants (HAPs), volatile organic compounds (VOCs), and odorous emissions.

- Sulphur oxides (SO_x) – particularly sulfur dioxide, a chemical compound with the formula SO_2. SO_2 is produced by volcanoes and in various industrial processes. Coal and petroleum often contain sulphur compounds, and their combustion generates sulphur dioxide. Further oxidation of SO_2, usually in the presence of a catalyst such as NO_2, forms H_2SO_4, and thus acid rain. This is one of the causes for concern over the environmental impact of the use of these fuels as power sources.

- Nitrogen oxides (NO_x) – Nitrogen oxides, particularly nitrogen dioxide, are expelled from high temperature combustion, and are also produced during thunderstorms by electric discharge. They can be seen as a brown haze dome above or a plume downwind of cities. Nitrogen dioxide is a chemical compound with the formula NO_2. It is one of several nitrogen oxides. One of the most prominent air pollutants, this reddish-brown toxic gas has a characteristic sharp, biting odor.

- Carbon monoxide (CO) – CO is a colorless, odorless, toxic yet non-irritating gas. It is a product of combustion of fuel such as natural gas, coal or wood. Vehicular exhaust contributes to the majority of carbon monoxide let into our atmosphere. It creates a smog type formation in the air that has been linked to many lung diseases and disruptions to the natural environment and animals. In 2013, more than half of the carbon monoxide emitted into our atmosphere was from vehicle traffic and burning one gallon of gas will often emit over 20 pounds of carbon monoxide into the air.

- Volatile organic compounds (VOC) – VOCs are a well-known outdoor air pollutant. They are categorized as either methane (CH_4) or non-methane (NMVOCs). Methane is an extremely efficient greenhouse gas which contributes to enhanced global warming. Other hydrocarbon VOCs are also significant greenhouse gases because of their role in creating ozone and prolonging the life of methane in the atmosphere. This effect varies depending on local air quality. The aromatic NMVOCs benzene, toluene and xylene are suspected carcinogens and may lead to leukemia with prolonged exposure. 1,3-butadiene is another dangerous compound often associated with industrial use.

- Particulates, alternatively referred to as particulate matter (PM), atmospheric particulate matter, or fine particles, are tiny particles of solid or liquid suspended in a gas. In contrast, aerosol refers to combined particles and gas. Some particulates occur naturally, originating from volcanoes, dust storms, forest and grassland fires, living vegetation, and sea spray. Human activities, such as the burning of fossil fuels in vehicles, power plants and various industrial processes also generate significant amounts of aerosols. Averaged worldwide, anthropogenic aerosols—those made by human activities—currently account for approximately 10 percent of our atmosphere. Increased levels of fine particles in the air are linked to health hazards such as heart disease, altered lung function and lung cancer. Particulates are related to respiratory infections and can be particularly harmful to those already suffering from conditions like asthma.

- Persistent free radicals connected to airborne fine particles are linked to cardiopulmonary disease.

- Toxic metals, such as lead and mercury, especially their compounds.

- Chlorofluorocarbons (CFCs) – harmful to the ozone layer; emitted from products are currently banned from use. These are gases which are released from air conditioners, refrigerators, aerosol sprays, etc. On release into the air, CFCs rise to the stratosphere. Here they come in contact with other gases and damage the ozone layer. This allows harmful ultraviolet rays to reach the earth's surface. This can lead to skin cancer, eye disease and can even cause damage to plants.

- Ammonia (NH_3) – emitted from agricultural processes. Ammonia is a compound with the formula NH_3. It is normally encountered as a gas with a characteristic pungent odor. Ammonia contributes significantly to the nutritional needs of terrestrial organisms by serving as a precursor to foodstuffs and fertilizers. Ammonia, either directly or indirectly, is also a building block for the synthesis of many pharmaceuticals. Although in wide use, ammonia is both caustic and hazardous. In the atmosphere, ammonia reacts with oxides of nitrogen and sulfur to form secondary particles.

- Odours — such as from garbage, sewage, and industrial processes.

- Radioactive pollutants – produced by nuclear explosions, nuclear events, war explosives, and natural processes such as the radioactive decay of radon.

Natural Air Pollution

When we think of pollution, we tend to think it's a problem that humans cause through ignorance or stupidity—and that's certainly true, some of the time. However, it's important to remember that some kinds of air pollution are produced naturally. Forest fires, erupting volcanoes, and gases released from radioactive decay of rocks inside

Earth are just three examples of natural air pollution that can have hugely disruptive effects on people and the planet.

Forest fires are one completely natural cause of air pollution. We'll never be able to prevent them breaking out or stop the pollution they cause; our best hope is to manage forests, where we can, so fires don't spread.

Forest fires (which often start naturally) can produce huge swathes of smoke that drift for miles over neighboring cities, countries, or continents. Giant volcanic eruptions can spew so much dust into the atmosphere that they block out significant amounts of sunlight and cause the entire planet to cool down for a year or more. Radioactive rocks can release a gas called radon when they decay, which can build up in the basements of buildings with serious effects on people's health (each year, around 21,000 people die of lung cancer, due to radon gas in the United States).

All these things are examples of serious air pollution that happen without any help from humans; although we can adapt to natural air pollution, and try to reduce the disruption it causes, we can never stop it happening completely.

Causes of Air Pollution

1. Burning of Fossil Fuels: Sulfur dioxide emitted from the combustion of fossil fuels like coal, petroleum and other factory combustibles is one the major cause of air pollution. Pollution emitting from vehicles including trucks, jeeps, cars, trains, airplanes cause immense amount of pollution. We rely on them to fulfill our daily basic needs of transportation. But, there overuse is killing our environment as dangerous gases are polluting the environment. Carbon Monooxide caused by improper or incomplete combustion and generally emitted from vehicles is another major pollutant along with Nitrogen Oxides, that is produced from both natural and man made processes.

2. Agricultural activities: Ammonia is a very common by product from agriculture related activities and is one of the most hazardous gases in the atmosphere. Use of insecticides, pesticides and fertilizers in agricultural activities has grown quite a lot. They emit harmful chemicals into the air and can also cause water pollution.

3. Exhaust from factories and industries: Manufacturing industries release large amount of carbon monoxide, hydrocarbons, organic compounds, and chemicals into the air thereby depleting the quality of air. Manufacturing industries can be found at every corner of the earth and there is no area that has not been affected by it. Petroleum refineries also release hydrocarbons and various other chemicals that pollute the air and also cause land pollution.

4. Mining operations: Mining is a process wherein minerals below the earth are extracted using large equipments. During the process dust and chemicals are released in the air causing massive air pollution. This is one of the reason which is responsible for the deteriorating health conditions of workers and nearby residents.

5. Indoor air pollution: Household cleaning products, painting supplies emit toxic chemicals in the air and cause air pollution. Have you ever noticed that once you paint walls of your house, it creates some sort of smell which makes it literally impossible for you to breathe.

Suspended particulate matter popular by its acronym SPM, is another cause of pollution. Referring to the particles afloat in the air, SPM is usually caused by dust, combustion etc.

Effects of Air Pollution

1. Respiratory and heart problems: The effects of Air pollution are alarming. They are known to create several respiratory and heart conditions along with Cancer, among other threats to the body. Several millions are known to have died due to direct or indirect effects of Air pollution. Children in areas exposed to air pollutants are said to commonly suffer from pneumonia and asthma.

2. Global warming: Another direct effect is the immediate alterations that the world is witnessing due to Global warming. With increased temperatures world wide, increase in sea levels and melting of ice from colder regions and icebergs, displacement and loss of habitat have already signaled an impending disaster if actions for preservation and normalization aren't undertaken soon.

3. Acid Rain: Harmful gases like nitrogen oxides and sulfur oxides are released into the atmosphere during the burning of fossil fuels. When it rains, the water droplets combines with these air pollutants, becomes acidic and then falls on the ground in the form of acid rain. Acid rain can cause great damage to human, animals and crops.

4. Eutrophication: Eutrophication is a condition where high amount of nitrogen present in some pollutants gets developed on sea's surface and turns itself into algae and and adversely affect fish, plants and animal species. The green colored algae that is present on lakes and ponds is due to presence of this chemical only.

5. Effect on Wildlife: Just like humans, animals also face some devastating affects of air pollution. Toxic chemicals present in the air can force wildlife species to move to

new place and change their habitat. The toxic pollutants deposit over the surface of the water and can also affect sea animals.

6. Depletion of Ozone layer: Ozone exists in earth's stratosphere and is responsible for protecting humans from harmful ultraviolet (UV) rays. Earth's ozone layer is depleting due to the presence of chlorofluorocarbons, hydro chlorofluorocarbons in the atmosphere. As ozone layer will go thin, it will emit harmful rays back on earth and can cause skin and eye related problems. UV rays also have the capability to affect crops.

Health Impacts of Air Pollution

In 2012, air pollution caused premature deaths on average of 1 year in Europe, and was a significant risk factor for a number of pollution-related diseases, including respiratory infections, heart disease, COPD, stroke and lung cancer. The health effects caused by air pollution may include difficulty in breathing, wheezing, coughing, asthma and worsening of existing respiratory and cardiac conditions. These effects can result in increased medication use, increased doctor or emergency department visits, more hospital admissions and premature death. The human health effects of poor air quality are far reaching, but principally affect the body's respiratory system and the cardiovascular system. Individual reactions to air pollutants depend on the type of pollutant a person is exposed to, the degree of exposure, and the individual's health status and genetics. The most common sources of air pollution include particulates, ozone, nitrogen dioxide, and sulfur dioxide. Children aged less than five years that live in developing countries are the most vulnerable population in terms of total deaths attributable to indoor and outdoor air pollution.

Mortality

The World Health Organization estimated in 2014 that every year air pollution causes the premature death of some 7 million people worldwide. India has the highest death rate due to air pollution. India also has more deaths from asthma than any other nation according to the World Health Organization. In December 2013 air pollution was estimated to kill 500,000 people in China each year. There is a positive correlation between pneumonia-related deaths and air pollution from motor vehicle emissions.

Annual premature European deaths caused by air pollution are estimated at 430,000. An important cause of these deaths is nitrogen dioxide and other nitrogen oxides (NOx) emitted by road vehicles. In a 2015 consultation document the UK government disclosed that nitrogen dioxide is responsible for 23,500 premature UK deaths per annum. Across the European Union, air pollution is estimated to reduce life expectancy by almost nine months. Causes of deaths include strokes, heart disease, COPD, lung cancer, and lung infections.

Urban outdoor air pollution is estimated to cause 1.3 million deaths worldwide per year. Children are particularly at risk due to the immaturity of their respiratory organ systems.

The US EPA estimated in 2004 that a proposed set of changes in diesel engine technology (Tier 2) could result in 12,000 fewer premature mortalities, 15,000 fewer heart attacks, 6,000 fewer emergency department visits by children with asthma, and 8,900 fewer respiratory-related hospital admissions each year in the United States.

The US EPA has estimated that limiting ground-level ozone concentration to 65 parts per billion, would avert 1,700 to 5,100 premature deaths nationwide in 2020 compared with the 75-ppb standard. The agency projected the more protective standard would also prevent an additional 26,000 cases of aggravated asthma, and more than a million cases of missed work or school. Following this assessment, the EPA acted to protect public health by lowering the National Ambient Air Quality Standards (NAAQS) for ground-level ozone to 70 parts per billion (ppb).

A new economic study of the health impacts and associated costs of air pollution in the Los Angeles Basin and San Joaquin Valley of Southern California shows that more than 3,800 people die prematurely (approximately 14 years earlier than normal) each year because air pollution levels violate federal standards. The number of annual premature deaths is considerably higher than the fatalities related to auto collisions in the same area, which average fewer than 2,000 per year.

Diesel exhaust (DE) is a major contributor to combustion-derived particulate matter air pollution. In several human experimental studies, using a well-validated exposure chamber setup, DE has been linked to acute vascular dysfunction and increased thrombus formation.

The mechanisms linking air pollution to increased cardiovascular mortality are uncertain, but probably include pulmonary and systemic inflammation.

Cardiovascular Disease

A 2007 review of evidence found ambient air pollution exposure is a risk factor correlating with increased total mortality from cardiovascular events (range: 12% to 14% per 10 microg/m3 increase).

Air pollution is also emerging as a risk factor for stroke, particularly in developing countries where pollutant levels are highest. A 2007 study found that in women, air pollution is not associated with hemorrhagic but with ischemic stroke. Air pollution was also found to be associated with increased incidence and mortality from coronary stroke in a cohort study in 2011. Associations are believed to be causal and effects may be mediated by vasoconstriction, low-grade inflammation and atherosclerosis Other mechanisms such as autonomic nervous system imbalance have also been suggested.

Lung Disease

Research has demonstrated increased risk of developing asthma and COPD from increased exposure to traffic-related air pollution. Additionally, air pollution has been associated with increased hospitalization and mortality from asthma and COPD. Chronic obstructive pulmonary disease (COPD) includes diseases such as chronic bronchitis and emphysema.

A study conducted in 1960–1961 in the wake of the Great Smog of 1952 compared 293 London residents with 477 residents of Gloucester, Peterborough, and Norwich, three towns with low reported death rates from chronic bronchitis. All subjects were male postal truck drivers aged 40 to 59. Compared to the subjects from the outlying towns, the London subjects exhibited more severe respiratory symptoms (including cough, phlegm, and dyspnea), reduced lung function (FEV1 and peak flow rate), and increased sputum production and purulence. The differences were more pronounced for subjects aged 50 to 59. The study controlled for age and smoking habits, so concluded that air pollution was the most likely cause of the observed differences. More recent studies have shown that air pollution exposure from traffic reduces lung function development in children and lung function may be compromised by air pollution even at low concentrations. Air pollution exposure also cause lung cancer in non smokers.

It is believed that much like cystic fibrosis, by living in a more urban environment serious health hazards become more apparent. Studies have shown that in urban areas patients suffer mucus hypersecretion, lower levels of lung function, and more self-diagnosis of chronic bronchitis and emphysema.

Cancer

A review of evidence regarding whether ambient air pollution exposure is a risk factor for cancer in 2007 found solid data to conclude that long-term exposure to PM2.5 (fine particulates) increases the overall risk of non-accidental mortality by 6% per a 10 microg/m3 increase. Exposure to PM2.5 was also associated with an increased risk of mortality from lung cancer (range: 15% to 21% per 10 microg/m3 increase) and total cardiovascular mortality (range: 12% to 14% per a 10 microg/m3 increase). The review further noted that living close to busy traffic appears to be associated with elevated risks of these three outcomes – increase in lung cancer deaths, cardiovascular deaths, and overall non-accidental deaths. The reviewers also found suggestive evidence that exposure to PM2.5 is positively associated with mortality from coronary heart diseases and exposure to SO2 increases mortality from lung cancer, but the data was insufficient to provide solid conclusions. Another investigation showed that higher activity level increases deposition fraction of aerosol particles in human lung and recommended avoiding heavy activities like running in outdoor space at polluted areas.

In 2011, a large Danish epidemiological study found an increased risk of lung cancer for patients who lived in areas with high nitrogen oxide concentrations. In this study, the association was higher for non-smokers than smokers. An additional Danish study, also in 2011, likewise noted evidence of possible associations between air pollution and other forms of cancer, including cervical cancer and brain cancer.

Cancer is mainly the result of environmental factors.

In December 2015, medical scientists reported that cancer is overwhelmingly a result of environmental factors, and not largely down to bad luck. Maintaining a healthy weight, eating a healthy diet, minimizing alcohol and eliminating smoking reduces the risk of developing the disease, according to the researchers.

Children

In the United States, despite the passage of the Clean Air Act in 1970, in 2002 at least 146 million Americans were living in non-attainment areas—regions in which the concentration of certain air pollutants exceeded federal standards. These dangerous pollutants are known as the criteria pollutants, and include ozone, particulate matter, sulfur dioxide, nitrogen dioxide, carbon monoxide, and lead. Protective measures to ensure children's health are being taken in cities such as New Delhi, India where buses now use compressed natural gas to help eliminate the "pea-soup" smog. A recent study in Europe has found that exposure to ultrafine particles can increase blood pressure in children.

Infants

Ambient levels of air pollution have been associated with preterm birth and low birth weight. A 2014 WHO worldwide survey on maternal and perinatal health found a statistically significant association between low birth weights (LBW) and increased levels of exposure to PM2.5. Women in regions with greater than average PM2.5 levels had statistically significant higher odds of pregnancy resulting in a low-birth weight infant even when adjusted for country-related variables. The effect is thought to be from stimulating inflammation and increasing oxidative stress.

A study by the University of York found that in 2010 exposure to PM2.5 was strongly associated with 18% of preterm births globally, which was approximately 2.7 million premature births. The countries with the highest air pollution associated preterm births were in South and East Asia, the Middle East, North Africa, and West sub-Saharan Africa.

The source of PM 2.5 differs greatly by region. In South and East Asia, pregnant women are frequently exposed to indoor air pollution because of the wood and other biomass fuels used for cooking which are responsible for more than 80% of regional pollution. In the Middle East, North Africa and West sub-Saharan Africa, fine PM comes from natural sources, such as dust storms. The United States had an estimated 50,000 preterm births associated with exposure to PM2.5 in 2010.

A study performed by Wang, et al. between the years of 1988 and 1991 has found a correlation between Sulfur Dioxide (SO2) and total suspended particulates (TSP) and preterm births and low birth weights in Beijing. A group of 74,671 pregnant women, in four separate regions of Beijing, were monitored from early pregnancy to delivery along with daily air pollution levels of Sulfur Dioxide and TSP (along with other particulates). The estimated reduction in birth weight was 7.3 g for every 100 µg/m3 increase in SO2 and 6.9g for each 100 µg/m3 increase in TSP. These associations were statistically significant in both summer and winter, although, summer was greater. The proportion of low birth weight attributable to air pollution, was 13%. This is the largest attributable risk ever reported for the known risk factors of low birth weight. Coal stoves, which are in 97% of homes, are a major source of air pollution in this area.

Brauer et al. studied the relationship between air pollution and proximity to a highway with pregnancy outcomes in a Vancouver cohort of pregnant woman using addresses to estimate exposure during pregnancy. Exposure to NO, NO2, CO PM10 and PM2.5 were associated with infants born small for gestational age (SGA). Women living <50meters away from an expressway or highway were 26% more likely to give birth to a SGA infant.

"Clean" Areas

Even in the areas with relatively low levels of air pollution, public health effects can be significant and costly, since a large number of people breathe in such pollutants. A study published in 2017 found that even in areas of the U.S. where ozone and PM2.5 meet federal standards, Medicare recipients who are exposed to more air pollution have higher mortality rates. A 2005 scientific study for the British Columbia Lung Association showed that a small improvement in air quality (1% reduction of ambient PM2.5 and ozone concentrations) would produce $29 million in annual savings in the Metro Vancouver region in 2010. This finding is based on health valuation of lethal (death) and sub-lethal (illness) affects.

Central Nervous System

Data is accumulating that air pollution exposure also affects the central nervous system.

In a June 2014 study conducted by researchers at the University of Rochester Medical Center, , it was discovered that early exposure to air pollution causes the same damaging changes in the brain as autism and schizophrenia. The study also shows that air pollution also affected short-term memory, learning ability, and impulsivity. Lead researcher Professor Deborah Cory-Slechta said that "When we looked closely at the ventricles, we could see that the white matter that normally surrounds them hadn't fully developed. It appears that inflammation had damaged those brain cells and prevented that region of the brain from developing, and the ventricles simply expanded to fill the space. Our findings add to the growing body of evidence that air pollution may play a role in autism, as well as in other neurodevelopmental disorders." Air pollution has a more significant negative effect on males than on females.

In 2015, experimental studies reported the detection of significant episodic (situational) cognitive impairment from impurities in indoor air breathed by test subjects who were not informed about changes in the air quality. Researchers at the Harvard University and SUNY Upstate Medical University and Syracuse University measured the cognitive performance of 24 participants in three different controlled laboratory atmospheres that simulated those found in "conventional" and "green" buildings, as well as green buildings with enhanced ventilation. Performance was evaluated objectively using the widely used Strategic Management Simulation software simulation tool, which is a well-validated assessment test for executive decision-making in an unconstrained situation allowing initiative and improvisation. Significant deficits were observed in the performance scores achieved in increasing concentrations of either volatile organic compounds (VOCs) or carbon dioxide, while keeping other factors constant. The highest impurity levels reached are not uncommon in some classroom or office environments.

Agricultural Effects

In India in 2014, it was reported that air pollution by black carbon and ground level ozone had cut crop yields in the most affected areas by almost half in 2011 when compared to 1980 levels.

Economic Effects

Air pollution costs the world economy $5 trillion per year as a result of productivity losses and degraded quality of life, according to a joint study by the World Bank and the Institute for Health Metrics and Evaluation (IHME) at the University of Washington. These productivity losses are caused by deaths due to diseases caused by air pollution. One out of ten deaths in 2013 was caused by diseases associated with air pollution and the problem is getting worse. The problem is even more acute in the developing world. "Children under age 5 in lower-income countries are more than 60 times as likely to die from exposure to air pollution as children in high-income countries." The report states that additional economic losses caused by air pollution, including health costs and the adverse effect on agricultural and other productivity were not calculated in the report, and thus the actual costs to the world economy are far higher than $5 trillion.

Air Pollution and Air Movement

Local air quality typically varies over time because of the effect of weather patterns. For example, air pollutants are diluted and dispersed in a horizontal direction by prevailing winds, and they are dispersed in a vertical direction by atmospheric instability. Unstable atmospheric conditions occur when air masses move naturally in a vertical direction, thereby mixing and dispersing pollutants. When there is little or no vertical movement of air (stable conditions), pollutants can accumulate near the ground and cause temporary but acute episodes of air pollution. With regard to air quality, unstable atmospheric conditions are preferable to stable conditions.

The degree of atmospheric instability depends on the temperature gradient (i.e., the rate at which air temperature changes with altitude). In the troposphere (the lowest layer of the atmosphere, where most weather occurs), air temperatures normally decrease as altitude increases; the faster the rate of decrease, the more unstable the atmosphere. Under certain conditions, however, a temporary "temperature inversion" may occur, during which time the air temperature increases with increasing altitude, and the atmosphere is very stable. Temperature inversions prevent the upward mixing and dispersion of pollutants and are the major cause of air pollution episodes. Certain geographic conditions exacerbate the effect of inversions. For example, Los Angeles, situated on a plain on the Pacific coast of California and surrounded by mountains that block horizontal air motion, is particularly susceptible to the stagnation effects of inversions—hence the infamous Los Angeles smog. On the opposite coast of North America another metropolis, New York City, produces greater quantities of pollutants than does Los Angeles but has been spared major air pollution disasters—only because of favourable climatic and geographic circumstances. During the mid-20th century, governmental efforts to reduce air pollution increased substantially after several major inversions, such as one weeklong air pollution episode in London in 1952 that was directly blamed for more than 4,000 deaths.

Movement of air is caused by temperature or pressure differences and is eperienced as wind. Where there are differences of pressure between two places, a pressure gradient exists, across which air moves: from the high-pressure region to the low-pressure region. This movement of air however, does not follow the quickest straight-line path. In fact, the air moving from high to low pressure follows a spiralling route, outwards from high pressure and inwards towards low pressure. This is due to the rotation of the Earth beneath the moving air, which causes an apparent deflection of the wind to the right in the Northern Hemisphere, and to the left in the Southern Hemisphere. The deflection of air is caused by the Coriolis force. Consequently, air blows anticlockwise around a low-pressure centre (depression) and clockwise around a high-pressure centre (anticyclone) in the Northern Hemisphere. This situation is reversed in the Southern Hemisphere.

Wind caused by differences in temperature is known as convection or advection. In the atmosphere, convection and advection transfer heat energy from warmer regions to colder regions, either at the Earth surface or higher up in the atmosphere. Small-scale air movement of this nature is observed during the formation of sea and land breezes, due to temperature differences between seawater and land. At a much larger scale, temperature differences across the Earth generate the development of the major wind belts. Such wind belts, to some degree, define the climate zones of the world.

Air temperature is generally higher at ground level due to heating by the Sun, and decreases with increasing altitude. This vertical temperature difference creates a significant uplift of air, since warmer air nearer the surface is lighter than colder air above it. This vertical uplift of air can generate clouds and rain. Sometimes air from warmer regions of the world collides with air from colder regions. This air mass convergence occurs in the mid-latitudes, where the warm air is forced to rise above the colder air, generating fronts and depressions.

Types of Air Pollution

Ambient - Outdoor Air Pollution

Outdoor air is often referred to as ambient air.

Ambient air pollution arises from both natural and human-derived sources. Air pollution has likely had adverse health effects throughout history due to natural occurrences such as volcanic eruptions and wildfires. In the modern era, burning fossil fuels, electric power generation, home heating, and motor vehicle transport has greatly increased emissions and pollution exposure. The importance of ambient air pollution was first appreciated in the 20th century, when cars, trucks, and other vehicles created "smog," or photochemical pollution and when public health crises arose from periods of intense pollution such as the London "killer fog" in 1952.

The most common air pollutants of ambient air include:

- Particulate matter (PM10 and PM2.5)
- Ozone (O_3)
- Nitrogen dioxide (NO_2)
- Carbon monoxide (CO)
- Sulphur dioxide (SO_2)

Outdoor Pollution

The levels of outdoor air pollution reach their peak in developing countries, most of them from Asia. The air outside is polluted mainly from vehicle exhaust and emissions from industries. Several pollutants are mixed in the air and a large portion of the world population is regularly exposed to harmful air quality.

Outdoor Air Pollution

Sources of outdoor or ambient air pollution are varied and include both natural and man-made ones. Natural outdoor air pollution includes oxides of sulphur and nitrogen from volcanoes, oceans, biological decay, lightning strikes and forest fires, VOCs and pollen from plants, grasses and trees, and particulate matter from dust storms. Natural pollution is all around us all of the time. However, sometimes concentrations can increase dramatically, for example after a volcanic eruption, or at the beginning of the growing season.

Perhaps of more concern, given our ability to have greater control over its release to the atmosphere, are the man-made air pollutants, which can have a detrimental impact on ambient air quality. The most common source of man-made air pollution outdoors is the burning of fossil fuels, such as coal, oil and gas, in power stations, industries, homes and road vehicles. Depending on the nature of the fuel and the type of combustion process, pollutants released into the atmosphere from the burning of fossil fuels include nitrogen oxides, sulphur dioxide, carbon monoxide, particulate matter, lead and volatile organic compounds (VOCs). Other sources of these pollutants include forest burning, chemical, fertiliser and paper manufacture, and waste incineration. These pollutants are all called primary pollutants because they have direct sources. By contrast, there are no direct ground-level emissions of ozone to the atmosphere. Most ozone in the atmosphere near ground level is formed indirectly by the action of sunlight on VOCs in the presence of nitrogen dioxide, although a small amount comes from high up in the atmosphere where it forms naturally in the ozone layer. Ozone is a secondary ambient air pollutant.

Both primary and secondary pollutants are, to a greater or lesser extent, detrimental to human health, depending on their concentration in the air, and the sensitivity of the individual. Consequently, national and international legislation exists to regulate and control the amount of pollution emitted to the atmosphere, and to ensure that objectives for improving ambient air quality are achieved.

Household Air Pollution

It refers to the physical, chemical, and biological characteristics of air in the indoor environment within a home, building, or an institution or commercial facility. Indoor air pollution is a concern in the developed countries, where energy efficiency improvements sometimes make houses relatively airtight, reducing ventilation and raising pollutant levels. Indoor air problems can be subtle and do not always produce easily recognized impacts on health. Different conditions are responsible for indoor air pollution in the rural areas and the urban areas.

In the developing countries, it is the rural areas that face the greatest threat from indoor pollution, where some 3.5 billion people continue to rely on traditional fuels such as firewood, charcoal, and cowdung for cooking and heating. Concentrations of indoor pollutants in households that burn traditional fuels are alarming. Burning such fuels produces large amount of smoke and other air pollutants in the confined space of the home, resulting in high exposure. Women and children are the groups most vulnerable as they spend more time indoors and are exposed to the smoke. In 1992, the World Bank designated indoor air pollution in the developing countries as one of the four most critical global environmental problems. Daily averages of pollutant level emitted indoors often exceed current WHO guidelines and acceptable levels. Although many hundreds of separate chemical agents have been identified in the smoke from biofuels, the four most serious pollutants are particulates, carbon monoxide, polycyclic organic matter, and formaldehyde. Unfortunately, little monitoring has been done in rural and poor urban indoor environments in a manner that is statistically rigorous.

In urban areas, exposure to indoor air pollution has increased due to a variety of reasons, including the construction of more tightly sealed buildings, reduced ventilation, the use of synthetic materials for building and furnishing and the use of chemical products, pesticides, and household care products. Indoor air pollution can begin within the building or be drawn in from outdoors. Other than nitrogen dioxide, carbon monoxide, and lead, there are a number of other pollutants that affect the air quality in an enclosed space.

Volatile organic compounds originate mainly from solvents and chemicals. The main indoor sources are perfumes, hair sprays, furniture polish, glues, air fresheners, moth repellents, wood preservatives, and many other products used in the house. The main health effect is the imitation of the eye, nose and throat. In more severe cases there may be headaches, nausea and loss of coordination. In the long term, some of the pollutants are suspected to damage to the liver and other parts of the body.

Tobacco smoke generates a wide range of harmful chemicals and is known to cause cancer. It is well known that passive smoking causes a wide range of problems to the passive smoker (the person who is in the same room with a smoker and is not himself/herself a smoker) ranging from burning eyes, nose, and throat irritation to cancer, bronchitis, severe asthma, and a decrease in lung function.

Pesticides , if used carefully and the manufacturers, instructions followed carefully they do not cause too much harm to the indoor air.

Biological pollutants include pollen from plants, mite, hair from pets, fungi, parasites, and some bacteria. Most of them are allergens and can cause asthma, hay fever, and other allergic diseases.

Asbestos is the leading cause of indoor air pollution. Asbestos can be found in various materials used commonly in the automotive industry as well as home construction. They are most commonly found in coatings, paints, building materials, and ceiling and floor tiles.

You won't find asbestos as often as you used because newer products do not contain asbestos. However, if you have an old home that was constructed a long time ago, the risks for asbestos are much greater than that of a newer home. Asbestos has been banned in the US and is no longer being used.

Formaldehyde is another leading cause of indoor air pollution. It is no longer produced in the United States due to its ban in 1970 but can still be found in paints, sealants, and wood floors.

Radon which can be found underneath your home in various types of bedrock and other building materials, can also be a cause of indoor air pollution. Radon can get into the walls of your home and put both you and your family at risk.

Causes of Air Pollution

There are various factors causing air pollution; natural and anthropogenic, which contribute to the introduction of particulates and gases into the atmosphere.

Air pollution is the mixing of unwanted and harmful substances such as chemicals, dust, auto emissions, suspended particles, gases among others in our atmosphere. It can be of two types; indoor and outdoor air pollution. It is a serious threat to the health of living beings and the different ecosystems found in our environment. According to WHO, it was the cause of death of approximately 7 million people around the world in 2014.

Natural Causes

Human activity is a major cause of air pollution, much of
which results from industrial processes

Natural forms of pollution are those that result from naturally-occurring phenomena. This means they are caused by periodic activities that are not man-made or the result of human activity. What's more, these sources of pollution are subject to natural cycles, being more common under certain conditions and less common under others. Being part of Earth's natural climatic variations also means that they are sustainable over long periods of time.

Dust and Wildfires

In large areas of open land that have little to no vegetation, and are particularly dry due to a lack of precipitation, wind can naturally create dust storms. This particulate matter, when added to the air, can have a natural warming effect and can also be a health hazard for living creatures. Particulate matter, when scattered into regions that have natural vegetation, can also be a natural impediment to photosynthesis.

Wildfires are a natural occurrence in wooded areas when prolonged dry periods occur, generally as a result of season changes and a lack of precipitation. The smoke and carbon monoxide caused by these fires contribute to carbon levels in the atmosphere, which allows for greater warming by causing a Greenhouse Effect.

Animal and Vegetation

The Chiwaukum Fire in Washington State

Animal digestion (particularly by cattle) is another cause of natural air pollution, leading to the release of methane, another greenhouse gas. In some regions of the world, vegetation – such as black gum, poplar, oak, and willow trees – emits significant amounts of volatile organic compounds (VOCs) on warmer days. These react with primary anthropogenic pollutants – specifically nitrogen oxides, sulfur dioxide and carbon compounds – to produce low-lying seasonal hazes that are rich in ozone.

Volcanic Activity

Volcanic eruptions are a major source of natural air pollution. When an eruption occurs, it produces tremendous amounts of sulfuric, chlorine, and ash products, which are released into the atmosphere and can be picked up by winds to be dispersed over large areas. Additionally, compounds like sulfur dioxide and volcanic ash have been known to have a natural cooling effect, due to their ability to reflect solar radiation.

Anthropogenic Causes

But by far the greatest contributing to air pollution today are those that are a result of human impact – i.e. man-made causes. These are largely the result of human reliance on fossil fuels and heavy industry, but can also be due to the accumulation of waste, modern agriculture, and other man-made processes.

Fossil-Fuel Emissions

Emissions by vehicles are a major cause of anthropogenic air pollution

The combustion of fossil fuels like coal, petroleum and other factory combustibles is a major cause of air pollution. These are generally used in power plants, manufacturing facilities (factories) and waste incinerators, as well as furnaces and other types of fuel-burning heating devices. Providing air conditioning and other services also requires significant amounts of electricity, which in turn leads to more emissions.

According to the Union of Concerned Scientists, industry accounts for 21% of greenhouse gas emissions in the US, while electricity generation accounted for another 31%. Meanwhile, emissions caused by gasoline-burning vehicles – i.e. CO, CO^2, nitrogen oxides, particulates and water vapor – are also a significant source of air pollution.

A study conducted by the UCS in 2013 showed that transportation accounted for more than half of the carbon monoxide and nitrogen oxides, and almost a quarter of the hydrocarbons emitted into the air in the US. Globally, the situation is similar, with minor variations according to sector. According to the IPCC Fifth Assessment Report (2014), industry accounted for 21% of total greenhouse gas emissions, electricity and heat production for another 25%, and transportation accounted for 14%.

Agriculture and Animal Husbandry

Greenhouse gas emissions from agriculture (aka. the cultivation of crops and livestock) is created by a combination of factors, one is the production of methane by cattle. Another cause is deforestation, where the need for pastureland and growing fields requires the removal of trees that would otherwise sequester carbon and clean the air.

According to the IPCC Fifth Assessment Report, agriculture accounts for 24% of annual emissions. However, this estimate does not include the $CO2$ that ecosystems remove from the atmosphere by sequestering carbon in biomass, dead organic matter and soils, which offset approximately 20% of emissions from this sector.

Waste

Landfills are also known to generate methane, which is not only a major greenhouse gas, but also an asphyxiant and highly flammable and potentially hazardous if a land-fills grow unchecked. Population growth and urbanization have a proportional relation-ship with the production of waste, which in turn leads to greater demand for dumping grounds that are far removed from urban environments. These locations thus became a significant source of methane production.

For some time, environmental scientists have been aware that the Earth has several self-regulating mechanisms. When it comes to the Earth's atmosphere, these mecha-nisms allow for the sequestration of carbon and other pollutants, ensuring that the bal-ance of its ecosystem remains unaffected. Unfortunately, the growing impact humanity has had on the planet is threatening to permanently alter that balance.

Anthropogenic Causes

- Mining and smelting – emit into the air a variety of metals adsorbed on partic-ulate matter that is suspended in the air due to crushing & processing of min-eralogical deposits;

- Mine tailing disposal – due to their fine particulate nature (resulting after crushing and processing mineral ores) constitute a source of metals to ambient air which could be spread by the wind over large areas;

- Foundry activities – emit into the air a variety of metals absorbed on particulate matter that is suspended in the air due to processing of metallic raw materials (including the use of furnaces);

- Various industrial processes may emit both organic and inorganic contami-nants through accidental spills and leaks of stored chemicals or the handling and storage of chemicals – especially of volatile inorganic chemicals;

- Transportation – emits a series of air pollutants (gases – including carbon monoxide, sulfur oxides, and nitrogen oxides - and particulate matter) through the tailpipe gases due to internal combustion of various fuels (usually gasses such as oxides of carbons, of sulfur, of nitrogen, as well as organic chemicals as PAHs);

- Construction and Demolition activities – pollute the air with various construc-tion materials. Of special threat is the demolition of old buildings which may contain a series of banned chemicals such as PCBs, PBDEs, asbestos;

- Coal Power Plants – when burning coal this may emit a series of gases as well as particulate matter with metals (such as As, Pb, Hg) and organic compounds (especially PAHs);

- Heating of buildings – emits a series of gases and particulate matters due to burning fossil fuels;

- Waste Incineration – depending on waste composition, various toxic gases, and particulate matter is emitted into the atmosphere;

- Landfill disposal practices – usually generate methane due to the intensification of natural microbial decaying activity in the disposal area;

- Agriculture – pollute the air usually through emissions of ammonia gas and the application of pesticides/herbicides/insecticides which contain toxic volatile organic compounds;

- Control burning in forest and agriculture management – includes controlled burning that will emit gases and particulate matter;

- Military activities – may introduce toxic gases through practices and training;

- Smoking – emits a series of toxic chemicals including a series of organic and inorganic chemicals, some of which are carcinogenic;

- Storage and use of household products such as paint, sprays, varnish, etc that contains organic solvents which volatilize in the air (hence the smell we all feel while using them);

- Dry cleaned clothes - may retain and emit in the atmosphere small amounts of chlorinated solvents (such as PCE) or petroleum solvents that have been used by the dry cleaners; this could eventually create a health risk if the clothes returned from the dry cleaners are stored in enclosed indoor spaces.

References

- Davis, Devra (2002). When Smoke Ran Like Water: Tales of Environmental Deception and the Battle Against Pollution. Basic Books. ISBN 0-465-01521-2

- "Study links traffic pollution to thousands of deaths". The Guardian. London, UK: Guardian Media Group. 2008-04-15. Archived from the original on 20 April 2008. Retrieved 2008-04-15

- Qian, Di (June 29, 2017). "Air Pollution and Mortality in the Medicare Population". New England Journal of Medicine. 376(26): 2513–2522

- Davis, Devra (2002). When Smoke Ran Like Water: Tales of Environmental Deception and the Battle Against Pollution. Basic Books. ISBN 0-465-01521-2

- European Court of Justice, CURIA (2008). "PRESS RELEASE No 58/08 Judgment of the Court of Justice in Case C-237/07" (PDF). Retrieved 24 January 2015

- Gauderman, W (2007). "Effect of exposure to traffic on lung development from 10 to 18 years of age: a cohort study". The Lancet. 369 (9561): 571–77. doi:10.1016/S0140-6736(07)60037-3

- J. Sunyer (2001). "Urban air pollution and Chronic Obstructive Pulmonary disease: a review". European Respiratory Journal. 17(5): 1024–33. doi:10.1183/09031936.01.17510240. PMID 11488305

- "Bucknell tent death: Hannah Thomas-Jones died from carbon monoxide poisoning". BBC News. 17 January 2013. Retrieved 22 September 2015

Chapter 2

Air Pollutants and their Impacts

Air pollutants are the substances that cause pollution in air. These can be gaseous, liquid droplets or solid particles. The classification and major types of air pollutants such as hydrogen sulphide, oxides of nitrogen, sulphur dioxide, carbon monoxide, etc. and their impacts have been discussed in this chapter.

Air Pollutants

Any substance in air that could, in high enough concentration, harm animals, humans, vegetation, and/or materials. Such pollutants may be present as solid particles, liquid droplets, or gases. Air pollutants fall into two main groups: (1) those emitted from identifiable sources and, (2) those formed in the air by interaction between other pollutants. Over one hundred air pollutants have been identified, which include halogen compounds, nitrogen compounds, oxygen compounds, radioactive compounds, sulphur (sulfur) compounds, and volatile organic chemicals (VOC).

Some air pollutants are poisonous. Inhaling them can increase the chance you'll have health problems. People with heart or lung disease, older adults and children are at greater risk from air pollution. Air pollution isn't just outside - the air inside buildings can also be polluted and affect your health.

Air pollutants occur both outdoors or indoors, and can be natural or man-made. Outdoor air pollution, sometimes called ambient air pollution, occurs in both urbanand rural areas, although a different mix of air pollutants may be found in the countryside to that found in a city. Typical urban air pollutants from man-made activities include nitrogen oxides, carbon monoxide, sulphur dioxide, hydrocarbonsand particulate matter. All these pollutants are called primary pollutants because they are emitted directly into the atmosphere. Common sources of these primary pollutants include power station and industrial plants (sulphur dioxide), and road transport (carbon monoxide, particulate matter and nitrogen oxides). Ozone is a secondary pollutant, formed in the air as a result of chemical reactions. Whilst ozone does build up within cities on hot summer days, higher levels are usually found in the countryside, because of the special nature of the reactions involving the formation of ozone.

Classification of Pollutants

The variety of matter emitted into the atmosphere by natural and anthropogenic sources is so diverse that it is difficult to classify air pollutants neatly.

However, they are classified in different ways as follows:

On the Basis of Nature

Depending upon the nature of the pollutants and their interaction with environment process, the pollution caused by different agents can be classified into the following categories:

1) Pollution Caused by Solid Wastes: The solid wastes includes the pollutants like garbage, rubbish, ashes, large wastes formed due to demolition and construction processes, dead animals wastes, agricultural wastes, etc.

2) Pollution Caused by Liquid Wastes: Oxygen cycle is nicely operated in aquatic system maintaining ecological balance. That is, the dissolved oxygen is used by aquatic living organisms for their respiration and in return, these liberate carbon dioxide. Carbon dioxide molecules are again used by green plants and algae in the process of photosynthesis. During photosynthesis, oxygen is again liberated to water which remains in dissolved state. However, if some organic matter (food for bacteria) enters the water course, then bacteria oxidize these materials consuming oxygen from water. At such a condition, if the process of re-oxygenation is slower than the process of deoxygenation, then the river will be devoid of life sustaining dissolved oxygen and aquatic living organisms will die.

The most important source of organic pollutants is sewage which contains faecal matter, urine, kitchen washing and oil washings. Sewage also contains a large number of pathogenic and harmless bacteria. The strength of organic waste material of sewage is measured in terms of Bio-chemical Oxygen Demand (B.O.D). The value is expressed in terms of mg of oxygen per litre of waste for 5 days at 20C. If the volume of B.O.D. is below 1500mg per litre, the sewage is termed as weak waste, if it is 4000mg per litre, it is medium and above this value it is termed as strong waste. However, if liquid industrial wastes containing acids, alkalis and poisonous substances enter the river, the aquatic life is affected and self-purification system of water is impaired. Pesticides and herbicides which enter water may kill some organisms or accumulate in fishes which, when consumed by man, pass on the chemicals giving rise to cumulative poisoning.

3) Pollution Caused by Gaseous Wastes: The gaseous wastes include Carbon monoxide (CO), Sulphur dioxide (SO_2), Nitrogen dioxide (NO_2), Ozone (O_3) and smog gases (composed of a complex mixture of photochemical oxidation products of hydrocarbons.

These gases are more abundant in the atmosphere of industrial cities.

4) Pollution from Waste without Weight: This type of pollution is also known as pollution by energy waste; Wastes without weight may be of the following types:

i) Radio-active Substance: Despite of all possible precautions in the functioning and maintenance of nuclear reactors, it is seen that minute quantity of radio-active waste escapes out into the environment. From the mining operation of the uranium to the use and final disposal of wastes from the reactor, radio-active materials continuously escape out into the environment. Besides, a lot of radio-active wastes enter into environment during the nuclear tests.

ii) Heat: A large quantity of waste heat energy is dissipated into environment by the way of hot liquid streams or hot gases released by industries and automobiles.

iii) Noise: The unwanted sound is known as noise. This sources of noise for the general public are the machines in the industry, traffic noise, indiscriminate use of transistor, radios, public address systems, etc.

On the Basis of Decomposition

1) Non-Degradable Pollutants: These are not broken down by the natural processes like action of microbes. Most of these pollutants get accumulated in the environment and also get biologically magnified as these moves along the food chains in an under-composed state. These may also react with other compounds in the environment to produce toxins. These can be further sub-divided into two more classes:

i) Waste: e.g., glass, plastic, phenolics, aluminium cans, etc.

ii) Poisons: e.g., radio-active substances, pesticides, smog gases, heavy metals like mercury, lead and their salts.

2) Degradable Pollutants or Bio-degradable Pollutants: These are natural organic substances which can be decomposed, removed or consumed and thus, reduced to acceptable levels either by natural processes like biological or microbial action or by some engineered systems, like sewage treatment plants. The degradable pollutants can be further sub-divided into two categories:

i) Rapidly Degradable or Non-Persistent Pollutant: The degradation of these pollutants is very faster process. For example, the decomposition of sewage and wastes of animals and plants is a faster process. The domestic sewage can be rapidly decomposed by natural processes. However, the problems become complicated when the input into environment get exceeded of the decomposition or dispersal capacity.

ii) Slowly Degradable or Persistent Pollutant: The degradation of these pollutants, is a very slower process. It seems as if the amount of pollutant remains unchanged with time. For example, degradation of synthetic compounds and radio-active elements like Iodine 137, Strontium 90 or Plutonium 239 takes a longer period of time.

On the Basis of Origin

Primary Pollutant

Vehicles are a major contributor to primary pollutants, emitting the majority
of CO and NOx emissions in Canada

Primary pollutants are any type of pollutant emitted directly into the environment. They differ from secondary pollutants because secondary pollutants must *form* in the atmosphere, whereas primary pollutants do not. Primary pollutants can be emitted from many sources including cars, coal-fired power plants, natural gas power plants, biomassburning, natural forest fires, volcanoes, and many more.

The effects of primary pollutants are of concern as they can be harmful to humans, animals and plants. Their contribution to the formation of secondary pollutants is also concerning, as this is what causes harmful ground level ozone to form, along with different smogs, especially in densely populated cities such as Los Angeles.

The emission of primary pollutants has decreased considerably in the past years, due to improved regulations, technology and economic changes.

Types of primary pollutants include:

- Nitrogen oxides (NOx)

- Carbon monoxide (CO)

- Volatile organic compounds (VOCs)

- Sulfur oxides (SOx)

- Particulate matter (PM)

- Mercury

Certain pollutants may be both primary *and* secondary pollutants. NOx for example is emitted from vehicles and power plants, yet it can also form in the atmosphere from other chemicals.

Secondary Pollutant

Photochemical smog, consisting of various secondary pollutants, over Shanghai

Secondary pollutants are any type of pollutant that is formed in the atmosphere. These pollutants are not emitted directly from a source, such as vehicles or power plants, instead they form as a result of the pollutants emitted from these. Pollutants that are directly input into the environment are called primary pollutants.

Secondary pollutants are concerning as they take many forms and can be formed from many different compounds. The phenomena of photochemical smog seen over densely populated cities is a result of the interactions of primary pollutants with other molecules in the airsuch as molecular oxygen, water and hydrocarbons, which combine to form yellow clouds that are harmful to humans. Photochemical smog is made up of various secondary pollutants like ozone, peroxyacyl nitrates(PANs) and nitric acid.

Different types of secondary pollutants include:

- Ozone (O_3)
- Sulfuric acid and nitric acid (component of acid rain)
- Particulate matter
- Nitrogen dioxide (NO_2)
- Peroxyacyl nitrates (PANs)
- and more

These substances essentially "cook up" in the atmosphere, and are typically found downwind of primary emissionsdue to the time it takes to produce them. When primary pollutants cannot be dispersed due to inversion layers in the atmosphere, smog is formed over the area where they were produced, which is why smog is so prominent in warm, dense cities. Secondary pollutants are very sensitive to weather patterns.

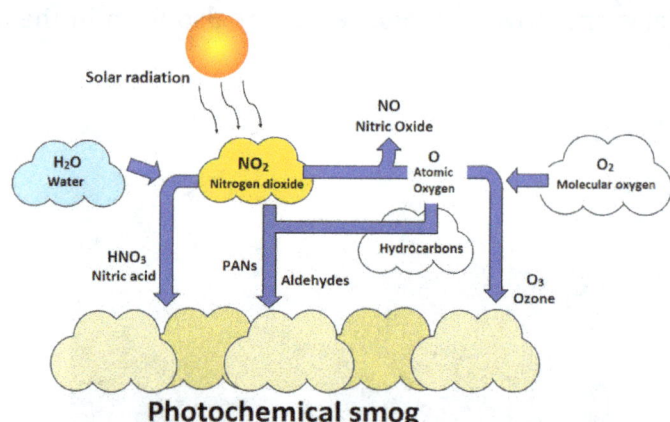

Photochemical smog formation; sunlight reacts with NO_2 which then interacts with other molecules in the air to form smog

Major Pollutants

Sulphur Dioxide

Sulfur dioxide is a gas released by both human and natural sources. It is a colorless gas with a pungent, irritating odor and taste. Sulfur dioxide is used in many industrial processes such as chemical preparation, refining, pulp-making and solvent extraction. In addition, it is used in the preparation and preservation of food due to its ability to prevent bacterial growth and browning of fruit.

It can be oxidized to sulphur trioxide, which in the presence of water vapour is readily transformed to sulphuric acid mist. SO_2 can be oxidized to form acid aerosols. SO_2 is a precursor to sulphates, which are one of the main components of respirable particles in the atmosphere.

Sulphur dioxide is an air pollutant that causes acid rain, haze and many health-related problems. It is produced predominantly when coal is burned to generate electricity.

Human Sources

Burning of fossil fuels such as coal, oil and natural gas are the main source of sulfur dioxide emissions. Coal fired power stations, in particular, are major sources of sulfur dioxide, with coal burning accounting for 50 percent of annual emissions, as explained by the Tropospheric Emission Monitoring Internet Service (TEMIS). Moreover, oil burning accounts for a further 25-30 percent. Sulfur dioxide emissions are released primarily as a result of generated electricity through fossil fuel burning power stations. Additional smaller sources of sulfur dioxide are released from industrial processes.

These include extracting metal from ore and the burning of fuels with a high sulfur content by locomotives, large ships and non-road equipment.

Natural Sources

Volcanic eruptions release large quantities of sulfur dioxide into the air. The vast quantities of sulfur dioxide released during one eruption can be enough to alter the global climate. Similarly, hot springs release sulfur dioxide into the atmosphere. Sulfur dioxide can even be produced by the reaction of hydrogen sulfide with the oxygen in the air. Hydrogen sulfide is released from marshes and regions in which biological decay is taking place.

As an Air Pollutant

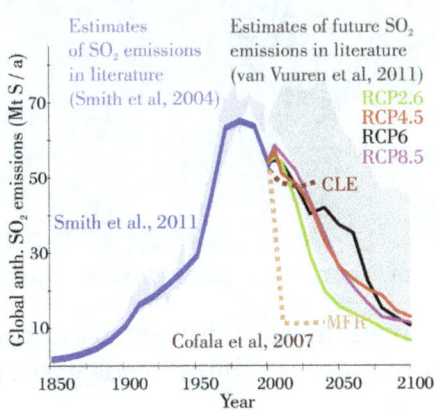

A sulfur dioxide plume from the Halema'uma'u vent, glows at night

Sulfur dioxide is a noticeable component in the atmosphere, especially following volcanic eruptions. According to the United States Environmental Protection Agency, the amount of sulfur dioxide released in the U.S. per year was:

A collection of estimates of past and future anthropogenic global sulphur dioxide emissions. The Cofala et al. estimates are for sensitivity studies on SO_2 emission policies, CLE: Current Legislation, MFR: Maximum Feasible Reductions. RCPs (Representative Concentration Pathways) are used in CMIP5 simulations for latest (2013–2014) IPCC 5th assessment report.

Year	SO$_2$
1970	31,161,000 short tons (28.3 Mt)
1980	25,905,000 short tons (23.5 Mt)
1990	23,678,000 short tons (21.5 Mt)
1996	18,859,000 short tons (17.1 Mt)
1997	19,363,000 short tons (17.6 Mt)
1998	19,491,000 short tons (17.7 Mt)
1999	18,867,000 short tons (17.1 Mt)

Sulfur dioxide is a major air pollutant and has significant impacts upon human health. In addition, the concentration of sulfur dioxide in the atmosphere can influence the habitat suitability for plant communities, as well as animal life. Sulfur dioxide emissions are a precursor to acid rain and atmospheric particulates. Due largely to the US EPA's Acid Rain Program, the U.S. has had a 33% decrease in emissions between 1983 and 2002. This improvement resulted in part from flue-gas desulfurization, a technology that enables SO$_2$ to be chemically bound in power plants burning sulfur-containing coal or oil. In particular, calcium oxide (lime) reacts with sulfur dioxide to form calcium sulfite:

$$CaO + SO_2 \rightarrow CaSO_3$$

Aerobic oxidation of the CaSO$_3$ gives CaSO$_4$, anhydrite. Most gypsum sold in Europe comes from flue-gas desulfurization.

Sulfur can be removed from coal during burning by using limestone as a bed material in fluidized bed combustion.

Sulfur can also be removed from fuels before burning, preventing formation of SO$_2$ when the fuel is burnt. The Claus process is used in refineries to produce sulfur as a byproduct. The Stretford process has also been used to remove sulfur from fuel. Redoxprocesses using iron oxides can also be used, for example, Lo-Cat or Sulferox.

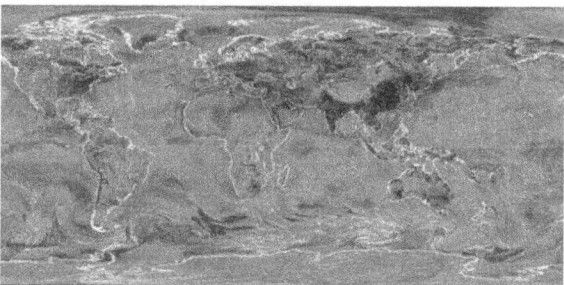

Emission of sulfur dioxide gases into the atmosphere

Fuel additives such as calcium additives and magnesium carboxylate may be used in marine engines to lower the emission of sulfur dioxide gases into the atmosphere.

As of 2006, China was the world's largest sulfur dioxide polluter, with 2005 emissions estimated to be 25,490,000 short tons (23.1 Mt). This amount represents a 27% increase since 2000, and is roughly comparable with U.S. emissions in 1980.

Safety

US Geological Survey volunteer tests for sulfur dioxide after the 2018 lower Puna eruption

Inhalation

Inhaling sulfur dioxide is associated with increased respiratory symptoms and disease, difficulty in breathing, and premature death. In 2008, the American Conference of Governmental Industrial Hygienists reduced the short-term exposure limit to 0.25 parts per million (ppm). The OSHA PEL is currently set at 5 ppm (13 mg/m³) time-weighted average. NIOSH has set the IDLHat 100 ppm. In 2010, the EPA "revised the primary SO_2 NAAQS by establishing a new one-hour standard at a level of 75 parts per billion (ppb). EPA revoked the two existing primary standards because they would not provide additional public health protection given a one-hour standard at 75 ppb."

A 2011 systematic review concluded that exposure to sulfur dioxide is associated with preterm birth.

Ingestion

In the United States, the Center for Science in the Public Interest lists the two food preservatives, sulfur dioxide and sodium bisulfite, as being safe for human consumption except for certain asthmatic individuals who may be sensitive to them, especially in large amounts. Symptoms of sensitivity to sulfiting agents, including sulfur dioxide, manifest as potentially life-threatening trouble breathing within minutes of ingestion.

Health Impacts of Sulfur Dioxide Emissions

Air pollution in the form of sulfur dioxide can have detrimental effects on human health. Such effects include breathing problems, particularly in asthmatics, whereas short-term exposure can lead to chest tightness and coughing and wheezing. Continued exposure to sulfur dioxide has been linked with alterations of the lungs defenses and aggravation of existing cardiovascular disease.

Environmental Impacts

The most common environmental impact of sulfur dioxide is the formation of acid rain. This occurs when the sulfur dioxide emissions combine with water vapor in the atmosphere, forming sulfuric acid, which falls to the ground as acid rain. Acid rain can acidify rivers and lakes, killing aquatic life in addition to damaging trees and plants. In addition, sulfur dioxide is a major precursor to particulate soot, which reduces air quality.

Hydrogen Sulphide

Hydrogen sulfide is a highly toxic and flammable gas. Because it is heavier than air it tends to accumulate at the bottom of poorly ventilated spaces. Although very pungent at first, it quickly deadens the sense of smell, so potential victims may be unaware of its presence until it is too late. H2S arises from virtually anywhere where elemental sulfur comes into contact with organic material, especially at high temperatures. Hydrogen sulfide is a covalent hydride chemically related to water (H_2O) since oxygen and sulfur occur in the same periodic table group. It often results when bacteria break down organic matter in the absence of oxygen, such as in swamps, and sewers (alongside the process of anaerobic digestion). It also occurs in volcanic gases, natural gas and some well waters. It is also important to note that Hydrogen sulfide is a central participant in the sulfur cycle, the biogeochemical cycle of sulfur on Earth. As mentioned above, sulfur-reducing and sulfate-reducing bacteria derive energy from oxidizing hydrogen or organic molecules in the absence of oxygen by reducing sulfur or sulfate to hydrogen sulfide. Other bacteria liberate hydrogen sulfide from sulfur-containing amino acids. Several groups of bacteria can use hydrogen sulfide as fuel, oxidizing it to elemental sulfur or to sulfate by using oxygen or nitrate as oxidant. The purple sulfur bacteria and the green sulfur bacteria use hydrogen sulfide as electron donor in photosynthesis, thereby producing elemental sulfur. (In fact, this mode of photosynthesis is older than the mode of cyanobacteria, algae and plants which uses wateras electron donor and liberates oxygen).

Hydrogen sulfide (H_2S) occurs naturally in crude petroleum, natural gas, volcanic gases, and hot springs. It can also result from bacterial breakdown of organic matter. It is also produced by human and animal wastes. Bacteria found in your mouth and gastrointestinal tract produce hydrogen sulfide from bacteria decomposing materials that contain vegetable or animal proteins. Hydrogen sulfide can also result from industrial activities, such as food processing, coke ovens, kraft paper mills, tanneries, and petroleum refineries. Hydrogen sulfide is a flammable, colorless gas with a characteristic odor of rotten eggs. It is commonly known as hydrosulfuric acid, sewer gas, and stink damp.

Impact on Safety (Short-Term)

Gas is a silent threat, often invisible to the body's senses. Inhalation is the primary route of exposure to hydrogen sulfide. Though it may be easily smelled by some people at small concentrations, continuous exposure to even low levels of H2S quickly deadens the sense of smell (olfactory desensitization). Exposure to high levels of the gas can deaden the sense of smell instantly. Although the scent of H2S is a characteristic, smell is not a dependable indicator of H2S gas presence or for indicating increasing concentrations of the gas.

H2S irritates the mucous membranes of the body and the respiratory tract, among other things. Following exposure, short-term, or acute, symptoms may include a headache, nausea, convulsions, and eye and skin irritation. Injury to the central nervous system can be immediate and serious after exposure. At high concentrations, only a few breaths are needed to induce unconsciousness, coma, respiratory paralysis, seizures, even death.

Impact on Health (Long-Term)

Those having prolonged exposure to high enough levels of H2S gas to cause unconsciousness may continue to experience headaches, reduced attention span and motor functions. Pulmonary effects of H2S gas exposure may not be apparent for up to 72 hours following removal from the affected environment. Delayed pulmonary edema, a buildup of excess fluid in the lungs, may also occur following exposure to high concentrations.

H2S does not accumulate in the body, but repeated/prolonged exposure to moderate levels can cause low blood pressure, headache, loss of appetite and weight loss. Prolonged exposure to low levels may cause painful skin rashes and irritated eyes. Repeated exposure over time to high levels of H2S may cause convulsions, coma, brain and heart damage, even death.

Impact on Facilities

Heavier than air, H2S gas accumulates in low lying areas of poorly ventilated spaces. In oil and gas applications, sour gas (products containing H2S gas) in the presence of air and moisture may form sulfuric acid, capable of corroding metals. Facility equipment, including the internal surfaces of various components, faces reduced durability and impact strength, potentially leading to premature failure.

Detection of H2s Gas

Hydrogen sulfide is a fast acting poison, impacting many systems within the body. Wearable gas sensors are necessary for early detection and alerting, as the body's senses are not reliable indicators. Importantly, gas detectors such as Blackline's G7 wireless gas detector, should be considered as they alert live monitoring personnel of worker H2S gas exposure. Devices with a fast response time and sturdy construction are important for use in harsh environments where H2S may occur. Additionally, as H2S may desensitize and render the body unconscious in no time at high concentrations, connected personal monitoring equipment is crucial.

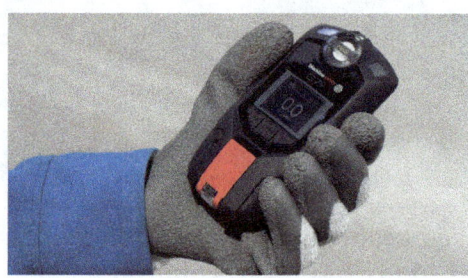

The Occupational Safety and Health Administration (OSHA) defines Permissible Exposure Limits (PELs) to H2S gas as follows:

- General Industry Ceiling Limit: 20 ppm

- General Industry Peak Limit: 50 ppm (up to 10 minutes if no other exposure during shift)

- Construction 8-hour Limit: 10 ppm

- Shipyard 8-hour limit: 10 ppm

Toxicity

Hydrogen sulfide is a broad-spectrum poison, meaning that it can poison several different systems in the body, although the nervous system is most affected. The toxicity of H_2S is comparable with that of carbon monoxide. It binds with iron in the mitochondrial cytochrome enzymes, thus preventing cellular respiration.

Since hydrogen sulfide occurs naturally in the body, the environment, and the gut, enzymes exist to detoxify it. At some threshold level, believed to average around 300–350 ppm, the oxidative enzymes become overwhelmed. Many personal safety gas detectors, such as those used by utility, sewage and petrochemical workers, are set to alarm at as low as 5 to 10 ppm and to go into high alarm at 15 ppm. Detoxification is effected by oxidation to sulfate, which is harmless. Hence, low levels of hydrogen sulfide may be tolerated indefinitely.

Diagnostic of extreme poisoning by H_2S is the discolouration of copper coins in the pockets of the victim. Treatment involves immediate inhalation of amyl nitrite, injections of sodium nitrite, or administration of 4-dimethylaminophenol in combination with inhalation of pure oxygen, administration of bronchodilators to overcome eventual bronchospasm, and in some cases hyperbaric oxygen therapy (HBOT). HBOT has clinical and anecdotal support.

Exposure to lower concentrations can result in eye irritation, a sore throat and cough, nausea, shortness of breath, and fluid in the lungs (pulmonary edema). These effects are believed to be due to the fact that hydrogen sulfide combines with alkali present in moist surface tissues to form sodium sulfide, a caustic. These symptoms usually go away in a few weeks.

Long-term, low-level exposure may result in fatigue, loss of appetite, headaches, irritability, poor memory, and dizziness. Chronic exposure to low level H_2S (around 2 ppm) has been implicated in increased miscarriage and reproductive health issues among Russian and Finnish wood pulp workers, but the reports have not (as of circa 1995) been replicated.

Short-term, high-level exposure can induce immediate collapse, with loss of breathing and a high probability of death. If death does not occur, high exposure to hydrogen sulfide can lead to cortical pseudolaminar necrosis, degeneration of the basal ganglia and cerebral edema. Although respiratory paralysis may be immediate, it can also be delayed up to 72 hours.

- 0.00047 ppm or 0.47 ppb is the odor threshold, the point at which 50% of a human panel can detect the presence of an odor without being able to identify it.

- 10 ppm is the OSHA permissible exposure limit (PEL) (8 hour time-weighted average).

- 10–20 ppm is the borderline concentration for eye irritation.

- 20 ppm is the acceptable ceiling concentration established by OSHA.

- 50 ppm is the acceptable maximum peak above the ceiling concentration for an 8-hour shift, with a maximum duration of 10 minutes.

- 50–100 ppm leads to eye damage.

- At 100–150 ppm the olfactory nerve is paralyzed after a few inhalations, and the sense of smell disappears, often together with awareness of danger.

- 320–530 ppm leads to pulmonary edema with the possibility of death.

- 530–1000 ppm causes strong stimulation of the central nervous system and rapid breathing, leading to loss of breathing.

- 800 ppm is the lethal concentration for 50% of humans for 5 minutes' exposure (LC50).

- Concentrations over 1000 ppm cause immediate collapse with loss of breathing, even after inhalation of a single breath.

Acute Exposure

Hydrogen sulfide's can cause inhibition of the cytochrome oxidase enzyme system resulting in lack of oxygen use in the cells. Anaerobic metabolism causes accumulation of lactic acid leading to an acid-base imbalance. The nervous system and cardiac tissues are particularly vulnerable to the disruption of oxidative metabolism and death is often the result of respiratory arrest. Hydrogen sulfide also irritates skin, eyes, mucous membranes, and the respiratory tract. Pulmonary effects may not be apparent for up to 72 hours after exposure.

Children do not always respond to chemicals in the same way that adults do. Different protocols for managing their care may be needed.

CNS

CNS injury is immediate and significant after exposure to hydrogen sulfide. At high concentrations, only a few breaths can lead to immediate loss of consciousness, coma, respiratory paralysis, seizures, and death. CNS stimulation may precede CNS depression. Stimulation manifests as excitation, rapid breathing, and headache; depression manifests as impaired gait, dizziness, and coma, possibly progressing to respiratory paralysis and death. In addition, decreased ability to smell hydrogen sulfide occurs at concentrations greater than 100 ppm.

Respiratory

Inhaled hydrogen sulfide initially affects the nose and throat. Low concentrations (50 ppm) can rapidly produce irritation of the nose, throat, and lower respiratory tract. Pulmonary manifestations include cough, shortness of breath, and bronchial or lung hemorrhage. Higher concentrations can provoke bronchitis and cause accumulation of fluid in the lungs, which may be immediate or delayed for up to 72 hours. Lack of oxygen may result in blue skin color.

Children may be more vulnerable to corrosive agents than adults because of the relatively smaller diameter of their airways. Children may also be more vulnerable to gas exposure because of increased minute ventilation per kg and failure to evacuate an area promptly when exposed.

Cardiovascular

High-dose exposures may cause insufficient cardiac output, irregular heartbeat, and conduction abnormalities.

Renal

Transient renal effects include blood, casts, and protein in the urine. Renal failure as a direct result of hydrogen sulfide toxicity has not been described, although it may occur secondary to cardiovascular compromise.

Gastrointestinal

Symptoms may include nausea and vomiting.

Dermal

Prolonged or massive exposure may cause burning, itching, redness, and painful inflammation of the skin. Exposure to the liquified gas can cause frostbite injury.

Ocular

Eye irritation may result in inflammation (i.e., keratoconjunctivitis) and clouding of the eye surface. Symptoms include blurred vision, sensitivity to light, and spasmodic blinking or involuntary closing of the eyelid.

Potential Sequelae

Inflammation of the bronchi can be a late development. Survivors of severe exposure may develop psychological disturbances and permanent damage to the brain and heart. The cornea may be permanently scarred.

Chronic Exposure

Hydrogen sulfide does not accumulate in the body. Nevertheless, repeated or prolonged exposure has been reported to cause low blood pressure, headache, nausea, loss of appetite, weight loss, ataxia, eye-membrane inflammation, and chronic cough. Neurologic symptoms, including psychological disorders, have been associated with chronic exposure. Chronic exposure may be more serious for children because of their potential longer latency period.

Carcinogenicity

Hydrogen sulfide has not been classified for carcinogenic effects.

Reproductive and Developmental Effects

There is some evidence to suggest that exposure to hydrogen Developmental Effects sulfide may be associated with an increased risk of spontaneous abortion. No information was located pertaining to placental transfer of hydrogen sulfide or to excretion of hydrogen sulfide in breast milk.

Oxides of Nitrogen

In atmospheric chemistry, NO_x is a generic term for the nitrogen oxides that are most relevant for air pollution, namely nitric oxide (NO) and nitrogen dioxide (NO_2). These gases contribute to the formation of smog and acid rain, as well as tropospheric ozone.

NO_x gases are usually produced from the reaction among nitrogen and oxygen during combustion of fuels, such as hydrocarbons, in air; especially at high temperatures, such as occur in car engines. In areas of high motor vehicle traffic, such as in large cities, the nitrogen oxides emitted can be a significant source of air pollution. NO_x gases are also produced naturally by lightning.

The term NO_x is chemistry shorthand for molecules containing one nitrogen and one or more oxygen atom. It is generally meant to include nitrous oxide (N_2O),although nitrous oxide is a fairly inert oxide of nitrogen that has many uses as an oxidizer for rockets and car engines, an anesthetic, and a propellant for aerosol spraysand whipped cream. Nitrous oxide plays hardly any role in air pollution, although it may have a significant impact on the ozone layer, and is a significant greenhouse gas.

NO_y (reactive, free radical) is defined as the sum of NO_x plus the NO_z compounds produced from the oxidation of NO_x which include nitric acid.

Sources of NOx Pollution

NOx is produced from the reaction of nitrogen and oxygen gases in the air during combustion, especially at high temperatures. In areas of high motor vehicle traffic, such as in large cities, the amount of nitrogen oxides emitted into the atmosphere as air pollution can be significant. NOx gases are formed whenever combustion occurs in the presence of nitrogen – e.g. in car engines; they are also produced naturally by lightning.

NOx Emissions in the EU -Share of Emissions by Sector Group, 2011

The pie chart below shows that road transport and energy production are the greatest sources of NOx emissions in the EU during 2011.

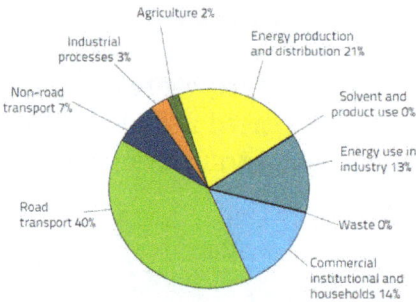

European Union emission inventory report 1990–2011 under the UNECE Convention on Long-range Trans-boundary Air Pollution (LRTAP)

Health Issues Created by NOx

NOx mainly impacts on respiratory conditions causing inflammation of the airways at high levels. Long term exposure can decrease lung function, increase the risk of respiratory conditions and increases the response to allergens. NOx also contributes to the formation of fine particles (PM) and ground level ozone, both of which are associated with adverse health effects.

The Impact of Nitrogen Dioxide on Ecosystems

High levels of NOx can have a negative effect on vegetation, including leaf damage and reduced growth. It can make vegetation more susceptible to disease and frost damage. A study of the effect of nitrogen dioxide and ammonia (NH_3) on the habitat of Epping Forest has revealed that pollution is likely to be significantly influencing ecosystem health in the forest. The study demonstrated that local traffic emissions contribute substantially to exceeding the critical levels and critical loads in the area. The critical level for the protection of vegetation is 30 µg/m3 measured as an annual average.

NOx also reacts with other pollutants in the presence of sunlight to form ozone which can damage vegetation at high concentrations.

Carbon Monoxide

Carbon monoxide is a colourless and odourless gas. It is found in smoke and is formed from the incomplete combustion of fuels such as peat, wood, coal, charcoal, natural gas, petrol, kerosene, oil, or propane.

Carbon monoxide is also found in exhaust fumes from cars, petrol and gas engines, gas ovens and cooktops, generators, lanterns, BBQ's and gas and wood heaters.

Effects of Carbon Monoxide Exposure on Health

When breathed in, carbon monoxide displaces oxygen in the blood and deprives the heart, brain and other vital organs of oxygen. Carbon monoxide may cause "flu-like" symptoms such as headache and tiredness, progressing to dizziness, confusion, nausea or fainting. Very high amounts of carbon monoxide in the body may result in oxygen deprivation, leading to loss of consciousness or death.

Whether someone develops health effects from exposure to carbon monoxide depends on a number of factors including:

- the levels of carbon monoxide in the environment (from smoke and also other environmental sources)
- how long a person is exposed
- a person's individual susceptibility, for example, having an existing heart or lung condition; having anaemia; being young, elderly or pregnant (the unborn child)

- the level of exercise or physical activity, which increases the amount of air breathed into the lungs (ie breathing rate)

- other lifestyle factors such as being a smoker.

Carbon monoxide levels have been consistently below standards in recent years and existing controls appear to be adequate.

Carbon Monoxide Air Quality Categories

The carbon monoxide (CO) data on our website will be shown in different colours, depending on the amount of CO in the air. The categories range from green – when levels of CO are low and air quality is very good – through to black – when high levels of CO result in very poor air quality.

Air quality category	CO ppm
Very good	0–2.9
Good	3.0–5.8
Fair	5.9–8.9
Poor	9.0–13.4
Very poor	13.5 or greater

Toxicity

Carbon monoxide poisoning is the most common type of fatal air poisoning in many countries. Carbon monoxide is colorless, odorless, and tasteless, but highly toxic. It combines with hemoglobin to produce carboxyhemoglobin, which usurps the space in hemoglobin that normally carries oxygen, but is ineffective for delivering oxygen to bodily tissues. Concentrations as low as 667 ppm may cause up to 50% of the body's hemoglobin to convert to carboxyhemoglobin. A level of 50% carboxyhemoglobin may result in seizure, coma, and fatality. In the United States, the OSHA limits long-term workplace exposure levels above 50 ppm.

The most common symptoms of carbon monoxide poisoning may resemble other types of poisonings and infections, including symptoms such as headache, nausea, vomiting, dizziness, fatigue, and a feeling of weakness. Affected families often believe they are victims of food poisoning. Infants may be irritable and feed poorly. Neurological signs include confusion, disorientation, visual disturbance, syncope (fainting), and seizures.

Some descriptions of carbon monoxide poisoning include retinal hemorrhages, and an abnormal cherry-red blood hue. In most clinical diagnoses these signs are seldom noticed. One difficulty with the usefulness of this cherry-red effect is that it corrects, or masks, what would otherwise be an unhealthy appearance, since the chief effect of removing deoxygenated hemoglobin is to make an asphyxiated person appear more normal, or a dead person appear more lifelike, similar to the effect of red colorants in embalming fluid. The "false" or unphysiologic red-coloring effect in anoxic CO-poisoned tissue is related to the meat-coloring commercial use of carbon monoxide.

Carbon monoxide also binds to other molecules such as myoglobin and mitochondrial cytochrome oxidase. Exposures to carbon monoxide may cause significant damage to the heart and central nervous system, especially to the globus pallidus, often with long-term chronic pathological conditions. Carbon monoxide may have severe adverse effects on the fetus of a pregnant woman.

Urban Pollution

Carbon monoxide is a temporary atmospheric pollutant in some urban areas, chiefly from the exhaust of internal combustion engines (including vehicles, portable and back-up generators, lawn mowers, power washers, etc.), but also from incomplete combustion of various other fuels (including wood, coal, charcoal, oil, paraffin, propane, natural gas, and trash).

Large CO pollution events can be observed from space over cities.

Role in Ground-Level Ozone Formation

Carbon monoxide is, along with aldehydes, part of the series of cycles of chemical reactions that form photochemical smog. It reacts with hydroxyl radical ($\cdot$OH) to produce a radical intermediate $\cdot$HOCO, which transfers rapidly its radical hydrogen to O_2 to form peroxy radical ($HO_2\cdot$) and carbon dioxide (CO_2). Peroxy radical subsequently reacts with nitrogen oxide (NO) to form nitrogen dioxide (NO_2) and hydroxyl radical. NO_2 gives $O(^3P)$ via photolysis, thereby forming O_3 following reaction with O_2. Since hydroxyl radical is formed during the formation of NO_2, the balance of the sequence of chemical reactions starting with carbon monoxide and leading to the formation of ozone is:

$$CO + 2O_2 + h\nu \rightarrow CO_2 + O_3$$

(where hν refers to the photon of light absorbed by the NO_2 molecule in the sequence)

Although the creation of NO_2 is the critical step leading to low level ozone formation, it also increases this ozone in another, somewhat mutually exclusive way, by reducing the quantity of NO that is available to react with ozone.

Indoor Pollution

In closed environments, the concentration of carbon monoxide can easily rise to lethal levels. On average, 170 people in the United States die every year from carbon monoxide produced by non-automotive consumer products. However, according to the Florida Department of Health, "every year more than 500 Americans die from accidental exposure to carbon monoxide and thousands more across the U.S. require emergency medical care for non-fatal carbon monoxide poisoning" These products include malfunctioning fuel-burning appliances such as furnaces, ranges, water heaters, and gas and kerosene room heaters; engine-powered equipment such as portable generators; fireplaces; and charcoal that is burned in homes and other enclosed areas. The American Association of Poison Control Centers (AAPCC) reported 15,769 cases of carbon monoxide poisoning resulting in 39 deaths in 2007. In 2005, the CPSC reported 94 generator-related carbon monoxide poisoning deaths. Forty-seven of these deaths were known to have occurred during power outages due to severe weather, including Hurricane Katrina. Still others die from carbon monoxide produced by non-consumer products, such as cars left running in attached garages. The Centers for Disease Control and Prevention estimates that several thousand people go to hospital emergency rooms every year to be treated for carbon monoxide poisoning.

Ozone

Without ozone, life on Earth would not have evolved in the way it has. The first stage of single cell organism development requires an oxygen-free environment. This type of environment existed on earth over 3000 million years ago. As the primitive forms of plant life multiplied and evolved, they began to release minute amounts of oxygen through the photosynthesis reaction (which converts carbon dioxide into oxygen). The build up of oxygen in the atmosphere led to the formation of the ozone layer in the upper atmosphere or stratosphere. This layer filters out incoming radiation in the "cell-damaging" ultraviolet (UV) part of the spectrum. Thus with the development of the ozone layer came the formation of more advanced life forms.

Ozone is a form of oxygen. The oxygen we breathe is in the form of oxygen molecules (O_2) - two atoms of oxygen bound together. Normal oxygen which we breathe is colourless and odourless. Ozone, on the other hand, consists of three atoms of oxygen bound together (O_3). Most of the atmosphere's ozone occurs in the region called the stratosphere. Ozone is colourless and has a very harsh odour. Ozone is much less common than normal oxygen. Out of 10 million air molecules, about 2 million are normal oxygen, but only 3 are ozone.

Most ozone is produced naturally in the upper atmosphere or stratosphere. While ozone can be found through the entire atmosphere, the greatest concentration occurs at altitudes between 19 and 30 km above the Earth's surface. This band of ozone-rich air is known as the "ozone layer". Ozone also occurs in very small amounts in the lowest few kilometres of the atmosphere, a region known as the troposphere. It is produced at ground level through a reaction between sunlight and volatile organic compounds (VOCs) and nitrogen oxides (NOx), some of which are produced by human activities such as driving cars. Ground-level ozone is a component of urban smog and can be harmful to human health.

Even though both types of ozone contain the same molecules, their presence in different parts of the atmosphere has very different consequences. Stratospheric ozone blocks harmful solar radiation - all life on Earth has adapted to this filtered solar radiation. Ground-level ozone, in contrast, is simply a pollutant. It will absorb some incoming solar radiation, but it cannot make up for ozone losses in the stratosphere.

Ozone Air Pollution

Red Alder leaf, showing discolouration caused by ozone pollution

Signboard in Gulfton, Houstonindicating an ozone watch

Ozone precursors are a group of pollutants, predominantly those emitted during the combustion of fossil fuels. Ground-level ozone pollution (tropospheric ozone) is created near the Earth's surface by the action of daylight UV rays on these precursors. The ozone at ground level is primarily from fossil fuel precursors, but methane is a natural precursor, and the very low natural background level of ozone at ground level is considered safe. This section examines the health impacts of fossil fuel burning, which raises ground level ozone far above background levels.

There is a great deal of evidence to show that ground-level ozone can harm lung function and irritate the respiratory system. Exposure to ozone (and the pollutants that produce it) is linked to premature death, asthma, bronchitis, heart attack, and other cardiopulmonary problems.

Long-term exposure to ozone has been shown to increase risk of death from respiratory illness. A study of 450,000 people living in United States cities saw a significant correlation between ozone levels and respiratory illness over the 18-year follow-up period. The study revealed that people living in cities with high ozone levels, such as Houston or Los Angeles, had an over 30% increased risk of dying from lung disease.

Air quality guidelines such as those from the World Health Organization, the United States Environmental Protection Agency(EPA) and the European Union are based on detailed studies designed to identify the levels that can cause measurable ill health effects.

According to scientists with the US EPA, susceptible people can be adversely affected by ozone levels as low as 40 nmol/mol. In the EU, the current target value for ozone concentrations is 120 µg/m³ which is about 60 nmol/mol. This target applies to all member states in accordance with Directive 2008/50/EC. Ozone concentration is measured as a maximum daily mean of 8 hour averages and the target should not be exceeded on more than 25 calendar days per year, starting from January 2010. Whilst the directive requires in the future a strict compliance with 120 µg/m³ limit (i.e. mean ozone concentration not to be exceeded on any day of the year), there is no date set for this requirement and this is treated as a long-term objective.

In the USA, the Clean Air Act directs the EPA to set National Ambient Air Quality Standards for several pollutants, including ground-level ozone, and counties out of compliance with these standards are required to take steps to reduce their levels. In May 2008, under a court order, the EPA lowered its ozone standard from 80 nmol/mol to 75 nmol/mol. The move proved controversial, since the Agency's own scientists and advisory board had recommended lowering the standard to 60 nmol/mol. Many public health and environmental groups also supported the 60 nmol/mol standard, and the World Health Organization recommends 51 nmol/mol.

On January 7, 2010, the U.S. Environmental Protection Agency (EPA) announced proposed revisions to the National Ambient Air Quality Standard (NAAQS) for the pollutant ozone, the principal component of smog:

> EPA proposes that the level of the 8-hour primary standard, which was set at 0.075 µmol/mol in the 2008 final rule, should instead be set at a lower level within the range of 0.060 to 0.070 µmol/mol, to provide increased protection for children and other "at risk" populations against an array of O_3 – related adverse health effects that range from decreased lung function and increased respiratory symptoms to serious indicators of respiratory morbidity including emergency department visits and hospital admissions for respiratory causes, and possibly cardiovascular-related morbidity as well as total non- accidental and cardiopulmonary mortality.

On October 26, 2015, the EPA published a final rule with an effective date of December 28, 2015 that revised the 8-hour primary NAAQS from 0.075 ppm to 0.070 ppm.

The EPA has developed an Air Quality Index (AQI) to help explain air pollution levels to the general public. Under the current standards, eight-hour average ozone mole fractions of 85 to 104 nmol/mol are described as "unhealthy for sensitive groups", 105 nmol/mol to 124 nmol/mol as "unhealthy", and 125 nmol/mol to 404 nmol/mol as "very unhealthy".

Ozone can also be present in indoor air pollution, partly as a result of electronic equipment such as photocopiers. A connection has also been known to exist between the increased pollen, fungal spores, and ozone caused by thunderstorms and hospital admissions of asthma sufferers.

In the Victorian era, one British folk myth held that the smell of the sea was caused by ozone. In fact, the characteristic "smell of the sea" is caused by dimethyl sulfide, a chemical generated by phytoplankton. Victorian British folk considered the resulting smell "bracing".

Particulate Matter

Atmospheric particulate matter, better known as particulate matter or particulates or particle pollution are microscopic particles which are comprised of liquid or solid matter and suspended in the earth's atmosphere. Created by both natural and man-made causes, particulates impact the earth's climate, precipitation levels and can have substantial negative effects on human health. Particulates are the deadliest form of air pollution because of the ability for them to deeply penetrate the lungs and blood streams unfiltered. These particles vary greatly in size, composition, and origin.

EPA defines Particle Pollution or PM as,

"Particle pollution (also called particulate matter or PM) is the term for a mixture of solid particles and liquid droplets found in the air. Some particles, such as dust, dirt, soot, or smoke, are large or dark enough to be seen with the naked eye. Others are so small they can only be detected using an electron microscope."

Particulate matter is often divided into two main groups, based on their size:

1. Inhalable coarse particles: These particles range from 2.5 micrometers to 10 micrometers in diameter (PM10 – PM2.5).

2. Fine particles: These particles are found in smoke and haze with a size up to 2.5 μm (PM2.5).

While Inhalable coarse particle are found near roadways and dusty industries, fine particles can be directly emitted from sources such as forest fires, or they can form when gases emitted from power plants, industries and automobiles react in the air.

Causes of Particulate Matter

Natural Causes

- Volcanoes – Erupting volcanoes eject large quantities of particulates including volcanic ash and gases into the atmosphere, volcanic eruptions have been directly associated with climate change since studies began.

- Dust storms – Strong winds can pick up vast clouds of dust which in turn are dispersed into the atmosphere and can take years to return to the surface.

- Forest and grassland fires – Wood and grass smoke contain a complex mixture of particulates such as carbon monoxide and hydrogen cyanide, which are lifted into the air and rest in the atmosphere.

- Living vegetation – Vegetation that emits particles to the air, such as isoprene, methanol and spores. These particles can be carried upwards by the wind and add to the level of particulates in the atmosphere.

- Sea spray – Due to the large amounts of plastics that have broken down to Nano scale, particles and can be found in ocean water all over the world. These hazardous particles can be thrown into the air by strong sea spray.

- Tornado's and hurricanes – These powerful weather systems can pick up large quantities of resting dust and pollutants just from the countryside, let alone when they pass through cities and encounter cement dust and higher levels of overall pollutants.

Man Made Causes

- Coal Combustion – Coal burning is still used in the majority of countries to generate heat and supplying energy, the burning of coal directly increases the amount of carbon monoxide and other hazardous particles into the atmosphere.

- Oil Combustion – Used for fueling vehicles, which in turn emit a large number of exhaust fumes containing hazardous particulates all over the globe, in huge quantities. Due to the large amounts if these pollutants in cities, countless deaths are caused by particulates.

- Wood combustion – The burning of wood is a wide scale cause of particulates, used for many purposes such as heating and generating power, the combustion process sends many toxic cocktails of pollutants into the atmosphere, such as soot.

- Construction – Cement dust is a large portion of overall global pollutants, because of the dust's small particle size, it can hang around in the air for quite some time. The use of vehicles in the construction industry and other known pollutants makes the construction industries in dire need of reforming.

- Demolition – Huge amounts of dust are thrown into the air during even the smallest demolition project, these particles are picked up into the wind, and again due to the small size of said particles, they can stay airborne for a very long time.

- Road dust – Roads are covered in microscopic dust and pollutants which are sent airborne by the air pressure changes and wind caused when a car uses a road, this happens all over the planet.

- Power plants – Plants that burn fossil fuels for energy and even nuclear plants disperse particulates on a huge scale, vast plumes of smoke will be found at most power plants, dispersing hundreds of cubic feet of pollutants every hour.

- Industrial – Manufacturing plastics and other materials which create toxic fumes are dispersed into the oceans, air and land. Which in turn adds to the huge amount of man-made pollutants.

- Agricultural – Pesticides and other volatile chemicals are sent into the air via sprayers and liquid jets. Again the agricultural industries use a large number of vehicles running on fuels such as petrol and diesel, which all attribute to the level of air pollution.

- Livestock – The livestock industry creates a huge amount of particulates which are dispersed into the air, ground and oceans. Even the animals themselves have been found to disperse great quantities of methane into the atmosphere.

- Deforestation – Felling trees for various industries impacts the rate at which trees would naturally produce carbon dioxide worldwide, however recent trends of replanting and harvesting cycles have reduced this.

- Poor condition of anti-pollution technology – In recent years the world has made a strong effort to combat the high levels of lethal pollutants, but unfortunately, some countries have produced very minimal reductions in annual pollution output.

- Tobacco smoke – Hundreds of toxic chemicals are present in tobacco smoke, and due to the millions of smokers worldwide, this leads to further pollution.

Effects of Particulate Matter

Climate Effects

- Volcanic eruptions – These have been linked with changes in the earth's climate. For example, in the 1600s a volcanic eruption in Peru (Huaynaputina) is believed to have caused a devastating famine in Russia which resulted in nearly 2 million deaths.

- Eruption of Mount Pinatubo – The eruption of 1991, the second largest terrestrial eruption of the 20th century, led to a world-wide temperature reduction of 0.5 degrees Celsius which lasted several years.

- Weather – Particulates are thought to affect weather on a regional scale and have been linked to the failure of the Indian Monsoon. Due to suppression of levels of evaporation of water from the Indian Ocean.

- Drought – Aerosol haze and particulates are believed to be pushing tropical rainfall southward, leading to a number of droughts across the world. Droughts worldwide have been occurring much more often since recording began.

- Rainfall declines – A decline in Australian rainfall have led researchers to believe the increase of pollutants from Asia have shifted multi-latitude systems southward.

- Greenhouse Gasses – Our atmosphere's molecular make up has changed dramatically since the industrial revolution. The increase in global industries has led to a build-up of so called greenhouse gasses in the atmosphere, which prevent heat from escaping the planet leading to global warming.

- Global dimming – Reductions in the earths direct irradiance have led researchers to believe that the increase in particulates in the atmosphere has impacted this, global dimming also creates a cooling effect, counteracting the heating of the greenhouse gasses.

- Ocean acidification – Due to the higher levels of carbon dioxide released by human activity. An estimated 30-40% of carbon dioxide dissolves into the oceans, causing harmful effects to ocean life such as coral bleaching.

Health Effects

- Asthma – A rising rate of diagnoses have been linked to the increased levels of fine pollutants in countries worldwide, particularly in areas with higher pollution.

- Lung cancer – Fine particles that penetrate deep into the human respiratory system and attack the bronchi, affecting the health of the lungs and leading to cancerous growths.

- Cardiovascular disease – Numerous different particulates have drastic effects on the heart and its functions, again caused by the fine particulates that easily pass into the human system unfiltered.

- Premature delivery – Exposure to high levels of air pollutants has led to an increase in the amount of failed pregnancy's, especially in towns and cities with higher levels of pollution.

- Birth defects – Particulates pass through the mother and into the child at any point of the pregnancy and can lead to a wide range of birth defects.

- Premature death – Typically higher in regions with high levels of air pollutants and aerosols.

- Vascular inflammation – Caused by a plaque build-up in the arteries, directly caused by particulates inhaled.

- Atherosclerosis – Hardening of the arteries that reduces elasticity, leading to heart problems, also caused by plaque build-up.

- Radiation exposure – A large number of particulates are formed up of radioactive material such as uranium and thorium, which is then inhaled or finds its way into crops which in turn are consumed.

- One Million Deaths – Every year are associated directly to the air pollution caused by the coal industry alone.

- 5 Million Deaths – Every year are believed to be caused by particulates worldwide.

Vegetation Effects

- Mortality – Stomatal openings are clogged, leading to failures during the photosynthesis process.

References

- Owen, Lewis A.; Pickering, Kevin T (1997). An Introduction to Global Environmental Issues. Taylor & Francis. pp. 33–. ISBN 978-0-203-97400-1

- Taylor, J.A.; Simpson, R.W.; Jakeman, A.J. (1987). "A hybrid model for predicting the distribution of sulphur dioxide concentrations observed near elevated point sources". Ecological Modelling. 36 (3–4): 269–296. doi:10.1016/0304-3800(87)90071-8. ISSN 0304-3800

- Cartlidge, Edwin (18 August 2015). "Superconductivity record sparks wave of follow-up physics". Nature News. Retrieved 18 August 2015

- Kumaresan, Deepak; Wischer, Daniela; Stephenson, Jason; Hillebrand-Voiculescu, Alexandra; Murrell, J. Colin (2014). "Microbiology of Movile Cave—A Chemolithoautotrophic Ecosystem". Geomicrobiology Journal. 31 (3): 186–193. doi:10.1080/01490451.2013.839764. ISSN 0149-0451

- Foulkes, Charles Howard (2001) [First published Blackwood & Sons, 1934]. "Gas!" The story of the special brigade. Published by Naval & Military P. p. 105. ISBN 1-84342-088-0

- U. Schumann & H. Huntrieser (2007). "The global lightning-induced nitrogen oxides source" (PDF). Atmos. Chem. Phys. 7: 3823. Retrieved 2016-05-31

- Carol Potera (2008). "Air Pollution: Salt Mist Is the Right Seasoning for Ozone". Environ Health Perspect. 116 (7): A288. doi:10.1289/ehp.116-a288. PMC 2453175. PMID 18629329

- Richard, Pohanish (2012). Sittig's Handbook of Toxic and Hazardous Chemicals and Carcinogens (2 ed.). Elsevier. p. 572. ISBN 978-1-4377-7869-4. Retrieved 5 September 2015

- "Surface chemistry of phase-pure M1 MoVTeNb oxide during operation in selective oxidation of propane to acrylic acid" (PDF). Journal of Catalysis. 285 (1): 48–60. January 2012

- Verma, A; Hirsch, D.; Glatt, C.; Ronnett, G.; Snyder, S. (1993). "Carbon monoxide: A putative neural messenger". Science. 259 (5093): 381–4. Bibcode:1993Sci...259..381V. doi:10.1126/science.7678352. PMID 7678352

Chapter 3

Indoor Air Quality

Indoor air quality is the quality of air within and around human habitations. It can be adversely affected by the presence of gases, microbes, etc. The diverse topics elaborated in this chapter include ventilation, indoor bioaerosols and passive smoking, which will help in developing a better perspective about indoor air quality.

In the last several years, a growing body of scientific evidence has indicated that the air within buildings and homes can be more polluted than the outdoor air in even the largest and most industrialized cities. Other research indicates that people spend approximately 90 percent of their time indoors. Thus, for many people, the risks to health may be greater due to exposure to air pollution indoors than outdoors. Indoor pollution sources that release gases or particles into the air are the primary cause of indoor air quality problems in buildings. Inadequate ventilation can increase indoor pollutant levels by not bringing in enough outdoor air to dilute emissions from indoor sources such as oil and gas; building materials and furnishings as diverse as deteriorated, asbestos-containing insulation, wet or damp carpet, furniture made of certain pressed wood products; Products for cleaning and maintenance; central heating and cooling systems and humidification devices; and outdoor sources such as vehicle emissions, pesticides, construction and demolition dust, and radon.

The connection between the use of a building either as a workplace or as a dwelling and the appearance, in certain cases, of discomfort and symptoms that may be the very definition of an illness is a fact that can no longer be disputed. The main culprit is contamination of various kinds within the building, and this contamination is usually referred to as "poor quality of indoor air". The adverse effects due to poor air quality in closed spaces affect a considerable number of people, since it has been shown that urban dwellers spend between 58 and 78% of their time in an indoor environment which is contaminated to a greater or lesser degree. These problems have increased with the construction of buildings that are designed to be more airtight and that recycle air with a smaller proportion of new air from the outside in order to be more energy efficient. The fact that buildings that do not offer natural ventilation present risks of exposure to contaminants is now generally accepted.

The term indoor air is usually applied to nonindustrial indoor environments: office buildings, public buildings (schools, hospitals, theatres, restaurants, etc.) and private dwellings. Concentrations of contaminants in the indoor air of these structures are usually of the same order as those commonly found in outdoor air, and are much lower than those found in air in industrial premises, where relatively well-known standards are applied in order to assess air quality. Even so, many building occupants complain of the quality of the air they breathe and there is therefore a need to investigate the situation. Indoor air quality began to be referred to as a problem at the end of the 1960s, although the first studies did not appear until some ten years later.

Although it would seem logical to think that good air quality is based on the presence in the air of the necessary components in suitable proportions, in reality it is the user, through respiration, who is the best judge of its quality. This is because inhaled air is perceived perfectly through the senses, as human beings are sensitive to the olfactory and irritant effects of about half a million chemical compounds. Consequently, if the occupants of a building are as a whole satisfied with the air, it is said to be of high quality; if they are unsatisfied, it is of poor quality. Does this mean that it is possible to predict on the basis of its composition how the air will be perceived? Yes, but only in part. This method works well in industrial environments, where specific chemical compounds related to production are known, and their concentrations in the air are measured and compared with threshold limit values. But in nonindustrial buildings where there may be thousands of chemical substances in the air but in such low concentrations that they are, perhaps, thousands of times less than the limits set for industrial environments, the situation is different. In most of these cases information about the chemical composition of indoor air does not allow us to predict how the air will be perceived, since the combined effect of thousands of these contaminants, together with temperature and humidity, can produce air that is perceived as irritating, foul, or stale—that is, of poor quality. The situation is comparable to what happens with the detailed composition of an item of food and its taste: chemical analysis is inadequate to predict whether the food will taste good or bad. For this reason, when a ventilation system and its regular maintenance are being planned, an exhaustive chemical analysis of indoor air is rarely called for.

Another point of view is that people are considered the only sources of contamination in indoor air. This would certainly be true if we were dealing with building materials, furniture and ventilation systems as they were used 50 years ago, when bricks, wood and steel predominated. But with modern materials the situation has changed. All materials contaminate, some a little and others much, and together they contribute to a deterioration in the quality of indoor air.

The changes in a person's health due to poor indoor air quality can show up as a wide array of acute and chronic symptoms and in the form of a number of specific illnesses. These are illustrated in figure. Although poor indoor air quality results in fully developed illness in only a few cases, it can give rise to malaise, stress, absenteeism and loss of productivity (with concomitant increases in production costs); and allegations about problems related to the building can develop rapidly into conflict between the occupants, their employers and the owners of the buildings.

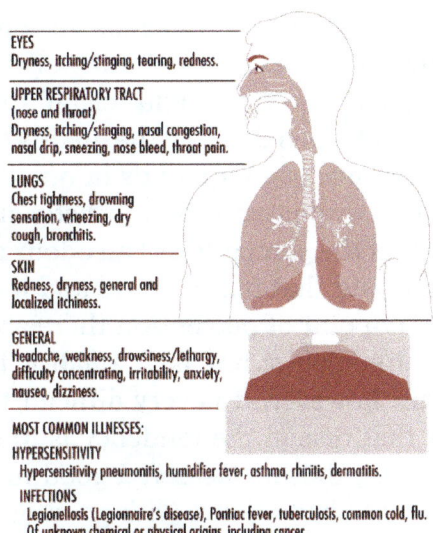

EYES
Dryness, itching/stinging, tearing, redness.

UPPER RESPIRATORY TRACT
(nose and throat)
Dryness, itching/stinging, nasal congestion,
nasal drip, sneezing, nose bleed, throat pain.

LUNGS
Chest tightness, drowning
sensation, wheezing, dry
cough, bronchitis.

SKIN
Redness, dryness, general and
localized itchiness.

GENERAL
Headache, weakness, drowsiness/lethargy,
difficulty concentrating, irritability, anxiety,
nausea, dizziness.

MOST COMMON ILLNESSES:
HYPERSENSITIVITY
    Hypersensitivity pneumonitis, humidifier fever, asthma, rhinitis, dermatitis.
INFECTIONS
    Legionellosis (Legionnaire's disease), Pontiac fever, tuberculosis, common cold, flu.
    Of unknown chemical or physical origins, including cancer.

Symptoms and illnesses related to the quality of indoor air

Normally it is difficult to establish precisely to what extent poor indoor air quality can harm health, since not enough information is available concerning the relationship between exposure and effect at the concentrations in which the contaminants are usually found. Hence, there is a need to take information obtained at high doses—as with exposures in industrial settings—and extrapolate to much lower doses with a corresponding margin of error. In addition, for many contaminants present in the air, the effects of acute exposure are well known, whereas there are considerable gaps in the data regarding both long-term exposures at low concentrations and mixtures of different contaminants. The concepts of no-effect-level (NOEL), harmful effect and tolerable effect, already confusing even in the sphere of industrial toxicology, are here even more difficult to define. There are few conclusive studies on the subject, whether relating to public buildings and offices or private dwellings.

Series of standards for outdoor air quality exist and are relied on to protect the general population. They have been obtained by measuring adverse effects on health resulting from exposure to contaminants in the environment. These standards are therefore useful as general guidelines for an acceptable quality of indoor air, as is the case with those proposed by the World Health Organization. Technical criteria such as the threshold limit value of the American Conference of Governmental Industrial Hygienists (ACGIH) in the United States and the limit values legally established for industrial environments in different countries have been set for the working, adult population and for specific lengths of exposure, and cannot therefore be applied directly to the general population. The American Society of Heating, Refrigeration and Air Conditioning Engineers (ASHRAE) in the United States has developed a series of standards and recommendations that are widely used in assessing indoor air quality.

Another aspect that should be considered as part of the quality of indoor air is its smell, because smell is often the parameter that ends up being the defining factor. The combination of a certain smell with the slight irritating effect of a compound in indoor air can lead us to define its quality as "fresh" and "clean" or as "stale" and "polluted". Smell is therefore very important when defining the quality of indoor air. While odours objectively depend on the presence of compounds in quantities above their olfactory thresholds, they are very often evaluated from a strictly subjective point of view. It should also be kept in mind that the perception of an odour may result from the smells of many different compounds and that temperature and humidity may also affect its characteristics. From the standpoint of perception there are four characteristics that allow us to define and measure odours: intensity, quality, tolerability and threshold. When considering indoor air, however, it is very difficult to "measure" odours from a chemical standpoint. For that reason the tendency is to eliminate odours that are "bad" and to use, in their place, those considered good in order to give air a pleasant quality. The attempt to mask bad odours with good ones usually ends in failure, because odours of very different qualities can be recognized separately and lead to unforeseeable results.

A phenomenon known as sick building syndrome occurs when more than 20% of the occupants of a building complain about air quality or have definite symptoms. It is evidenced by a variety of physical and environmental problems associated with non-industrial indoor environments. The most common features seen in cases of sick building syndrome are the following: those affected complain of non-specific symptoms similar to the common cold or respiratory illnesses; the buildings are efficient as regards energy conservation and are of modern design and construction or recently remodelled with new materials; and the occupants cannot control the temperature, humidity and illumination of the workplace.

The estimated percentage distribution of the most common causes of sick building syndrome are inadequate ventilation due to lack of maintenance; poor distribution and insufficient intake of fresh air (50 to 52%); contamination generated indoors, including from office machines, tobacco smoke and cleaning products (17 to 19%); contamination from the outside of the building due to inadequate placement of air intake and exhaust vents (11%); microbiological contamination from stagnant water in the ducts of the ventilation system, humidifiers and refrigeration towers (5%); and formaldehyde and other organic compounds emitted by building and decoration materials (3 to 4%). Thus, ventilation is cited as an important contributory factor in the majority of cases.

Another question of a different nature is that of building-related illnesses, which are less frequent, but often more serious, and are accompanied by very definite clinical signs and clear laboratory findings. Examples of building-related illnesses are hypersensitivity pneumonitis, humidifier fever, legionellosis and Pontiac fever. A fairly general opinion among investigators is that these conditions should be considered separately from sick building syndrome.

Studies have been done to ascertain both the causes of air quality problems and their possible solutions. In recent years, knowledge of the contaminants present in indoor air and the factors contributing to a decline in indoor air quality has increased considerably, although there is a long way to go. Studies carried out in the last 20 years have shown that the presence of contaminants in many indoor environments is higher than anticipated, and moreover, different contaminants have been identified from those that exist in outside air. This contradicts the assumption that indoor environments without industrial activity are relatively free of contaminants and that in the worst of cases they may reflect the composition of outside air. Contaminants such as radon and formaldehyde are identified almost exclusively in the indoor environment.

Indoor air quality, including that of dwellings, has become a question of environmental health in the same way as has happened with control of outdoor air quality and exposure at work. Although, as already mentioned, an urban person spends 58 to 78% of his or her time indoors, it should be remembered that the most susceptible persons, namely the elderly, small children and the sick, are the ones who spend most of their time indoors. This subject began to be particularly topical from around 1973 onwards, when, because of the energy crisis, efforts directed at energy conservation concentrated on reducing the entry of outside air into indoor spaces as much as possible in order to minimize the cost of heating and cooling buildings. Although not all the problems relating to indoor air quality are the result of actions aimed at saving energy, it is a fact that as this policy spread, complaints about indoor air quality began to increase, and all the problems appeared.

Another item requiring attention is the presence of micro-organisms in indoor air which can cause problems of both an infectious and an allergic nature. It should not be forgotten that micro-organisms are a normal and essential component of ecosystems. For example, saprophytic bacteria and fungi, which obtain their nutrition from dead organic material in the environment, are found normally in the soil and atmosphere, and their presence can also be detected indoors. In recent years problems of biological contamination in indoor environments have received considerable attention.

The outbreak of Legionnaire's disease in 1976 is the most discussed case of an illness caused by a micro-organism in the indoor environment. Other infectious agents, such as viruses that can cause acute respiratory illness, are detectable in indoor environments, especially if the occupation density is high and much recirculation of air is taking place. In fact, the extent to which micro-organisms or their components are implicated in the outbreak of building-associated conditions is not known. Protocols for demonstrating and analysing many types of microbial agents have been developed only to a limited degree, and in those cases where they are available, the interpretation of the results is sometimes inconsistent.

Origins of Contaminants

Indoor contamination has different origins: the occupants themselves; inadequate materials or materials with technical defects used in the construction of the building; the work performed within; excessive or improper use of normal products (pesticides, disinfectants, products used for cleaning and polishing); combustion gases (from smoking, kitchens, cafeterias and laboratories); and cross-contamination coming from other poorly ventilated zones which then diffuses towards neighbouring areas and affects them. It should be borne in mind that substances emitted in indoor air have much less opportunity of being diluted than those emitted in outdoor air, given the difference in the volumes of air available. As regards biological contamination, its origin is most frequently due to the presence of stagnant water, materials impregnated with water, exhausts and so on, and to defective maintenance of humidifiers and refrigeration towers.

Finally, contamination coming from outside must also be considered. As regards human activity, three main sources may be mentioned: combustion in stationary sources (power stations); combustion in moving sources (vehicles); and industrial processes. The five main contaminants emitted by these sources are carbon monoxide, oxides of sulphur, oxides of nitrogen, volatile organic compounds (including hydrocarbons), polycyclic aromatic hydrocarbons and particles. Internal combustion in vehicles is the principal source of carbon monoxide and hydrocarbons and is an important source of oxides of nitrogen. Combustion in stationary sources is the main origin of oxides of sulphur. Industrial processes and stationary sources of combustion generate more than half of the particles emitted into the air by human activity, and industrial processes can be a source of volatile organic compounds. There are also contaminants generated naturally that are propelled through the air, such as particles of volcanic dust, soil and sea salt, and spores and micro-organisms. The composition of outdoor air varies from place to place, depending both on the presence and the nature of the sources of contamination in the vicinity and on the direction of the prevailing wind. If there are no sources generating contaminants, the concentration of certain contaminants that will typically be found in "clean" outdoor air are as follows: carbon dioxide, 320 ppm; ozone, 0.02 ppm: carbon monoxide, 0.12 ppm; nitric oxide, 0.003 ppm; and nitrogen dioxide, 0.001 ppm. However, urban air always contains much higher concentrations of these contaminants.

Apart from the presence of the contaminants originating from outside, it sometimes happens that contaminated air from the building itself is expelled to the exterior and then returns inside again through the intakes of the air-conditioning system. Another possible way by which contaminants may enter from the exterior is by infiltration through the foundations of the building (e.g., radon, fuel vapours, sewer effluvia, fertilizers, insecticides and disinfectants). It has been shown that when the concentration of a contaminant in the outdoor air increases, its concentration in the air inside the building also increases, although more slowly (a corresponding relationship obtains when the concentration decreases); it is therefore said that buildings exert a shielding effect against external contaminants. However, the indoor environment is not, of course, an exact reflection of the conditions outside.

Contaminants present in indoor air are diluted in the outdoor air that enters the building and they accompany it when it leaves. When the concentration of a contaminant is less in the outdoor air than the indoor air, the interchange of indoor and outdoor air will result in a reduction in the concentration of the contaminant in the air inside the building. If a contaminant originates from outside and not inside, this interchange will result in a rise in its indoor concentration.

Models for the balance of amounts of contaminants in indoor air are based on the calculation of their accumulation, in units of mass versus time, from the difference between the quantity that enters plus what is generated indoors, and what leaves with the air plus what is eliminated by other means. If appropriate values are available for each of the factors in the equation, the indoor concentration can be estimated for a wide range of conditions. Use of this technique makes possible the comparison of different alternatives for controlling an indoor contamination problem.

Buildings with low interchange rates with outdoor air are classified as sealed or energy-efficient. They are energy-efficient because less cold air enters in winter, reducing the energy required to heat the air to the ambient temperature, thus cutting the cost of heating. When the weather is hot, less energy is also used to cool the air. If the building does not have this property, it is ventilated through open doors and windows by a process of natural ventilation. Although they may be closed, differences of pressure, resulting both from the wind and from the thermal gradient existing between the interior and the exterior, force the air to enter through crevices and cracks, window and door joints, chimneys and other apertures, giving rise to what is called ventilation by infiltration.

The ventilation of a building is measured in renewals per hour. One renewal per hour means that a volume of air equal to the volume of the building enters from outside every hour; in the same way, an equal volume of indoor air is expelled to the exterior every hour. If there is no forced ventilation (with a ventilator) this value is difficult to determine, although it is considered to vary between 0.2 and 2.0 renewals per hour. If the other parameters are assumed to be unchanged, the concentration of contaminants generated indoors will be less in buildings with high renewal values, although a high renewal value is not a complete guarantee of indoor air quality. Except in areas with marked atmospheric pollution, buildings that are more open will have a lower concentration of contaminants in the indoor air than those constructed in a more closed manner. However, buildings that are more open are less energy-efficient. The conflict between energy efficiency and air quality is of great importance.

Much action undertaken to reduce energy costs affects indoor air quality to a greater or lesser extent. In addition to reducing the speed with which the air circulates within the building, efforts to increase the insulation and waterproofing of the building involve the installation of materials that may be sources of indoor contamination. Other action, such as supplementing old and frequently inefficient central heating systems with secondary sources that heat or consume the indoor air can also raise contaminant levels in indoor air.

Contaminants whose presence in indoor air is most frequently mentioned, apart from those coming from outside, include metals, asbestos and other fibrous materials, formaldehyde, ozone, pesticides and organic compounds in general, radon, house dust and biological aerosols. Together with these, a wide variety of types of micro-organisms can be found, such as fungi, bacteria, viruses and protozoa. Of these, the saprophytic fungi and bacteria are relatively well known, probably because a technology is available for measuring them in air. The same is not true of agents such as viruses, rickettsiae, chlamydias, protozoa and many pathogenic fungi and bacteria, for the demonstration and counting of which no methodology is as yet available. Among the infectious agents, special mention should be made of: Legionella pneumophila, Mycobacterium avium, viruses, Coxiella burnetii and Histoplasma capsulatum; and among the allergens: Cladosporium, Penicillium and Cytophaga.

Common Pollutants

Second-hand Smoke

Second-hand smoke is tobacco smoke which affects people other than the 'active' smoker. Second-hand tobacco smoke includes both a gaseous and a particulate phase, with particular hazards arising from levels of carbon monoxide (as indicated below) and very small particulates (fine particular matter at especially PM2.5 size, and PM10) which get into the bronchioles and alveoles in the lung. The only certain method to improve indoor air quality as regards second-hand smoke is the implementation of comprehensive smoke-free laws.

Radon

Radon is an invisible, radioactive atomic gas that results from the radioactive decay of radium, which may be found in rock formations beneath buildings or in certain building materials themselves. Radon is probably the most pervasive serious hazard for indoor air in the United States and Europe, probably responsible for tens of thousands of deaths from lung cancer each year. There are relatively simple test kits for do-it-yourself radon gas testing, but if a home is for sale the testing must be done by a licensed person in some U.S. states. Radon gas enters buildings as a soil gas and is a heavy gas and thus will tend to accumulate at the lowest level. Radon may also be introduced into a building through drinking water particularly from bathroom showers. Building materials can be a rare source of radon, but little testing is carried out for stone, rock or tile products brought into building sites; radon accumulation is greatest for well insulated homes. The half life for radon is 3.8 days, indicating that once the source is removed, the hazard will be greatly reduced within a few weeks. Radon mitigation methods include sealing concrete slab floors, basement foundations, water drainage systems, or by increasing ventilation. They are usually cost effective and can greatly reduce or even eliminate the contamination and the associated health risks.

Molds and other Allergens

These biological chemicals can arise from a host of means, but there are two common classes: (a) moisture induced growth of mold colonies and (b) natural substances released into the air such as animal dander and plant pollen. Mold is always associated with moisture, and its growth can be inhibited by keeping humidity levels below 50%. Moisture buildup inside buildings may arise from water penetrating compromised areas of the building envelope or skin, from plumbing leaks, from condensationdue to improper ventilation, or from ground moisture penetrating a building part. Even something as simple as drying clothes indoors on radiators can increase the risk of exposure to (amongst other things) Aspergillus - a highly dangerous mould that can be fatal for asthma sufferers and the elderly. In areas where cellulosic materials (paper and wood, including drywall) become moist and fail to dry within 48 hours, mold mildew can propagate and release allergenic spores into the air.

In many cases, if materials have failed to dry out several days after the suspected water event, mold growth is suspected within wall cavities even if it is not immediately visible. Through a mold investigation, which may include destructive inspection, one should be able to determine the presence or absence of mold. In a situation where there is visible mold and the indoor air quality may have been compromised, mold remediation may be needed. Mold testing and inspections should be carried out by an independent investigator to avoid any conflict of interest and to insure accurate results; free mold testing offered by remediation companies is not recommended.

There are some varieties of mold that contain toxic compounds (mycotoxins). However, exposure to hazardous levels of mycotoxin via inhalation is not possible in most cases, as toxins are produced by the fungal body and are not at significant levels in the released spores. The primary hazard of mold growth, as it relates to indoor air quality, comes from the allergenic properties of the spore cell wall. More serious than most allergenic properties is the ability of mold to trigger episodes in persons that already have asthma, a serious respiratory disease.

Carbon Monoxide

One of the most acutely toxic indoor air contaminants is carbon monoxide (CO), a colourless, odourless gas that is a byproduct of incomplete combustion of fossil fuels. Common sources of carbon monoxide are tobacco smoke, space heaters using fossil fuels, defective central heating furnaces and automobile exhaust. By depriving the brain of oxygen, high levels of carbon monoxide can lead to nausea, unconsciousness and death. According to the American Conference of Governmental Industrial

Hygienists (ACGIH), the time-weighted average (TWA) limit for carbon monoxide (630-08-0) is 25 ppm.

Indoor levels of CO are systematically improving due to increasing implementation of smoke-free laws.

Volatile Organic Compounds

Volatile organic compounds (VOCs) are emitted as gases from certain solids or liquids. VOCs include a variety of chemicals, some of which may have short- and long-term adverse health effects. Concentrations of many VOCs are consistently higher indoors (up to ten times higher) than outdoors. VOCs are emitted by a wide array of products numbering in the thousands. Examples include: paints and lacquers, paint strippers, cleaning supplies, pesticides, building materials and furnishings, office equipment such as copiers and printers, correction fluids and carbonless copy paper, graphics and craft materials including glues and adhesives, permanent markers, and photographic solutions.

Chlorinated drinking water releases chloroform when hot water is used in the home. Benzene is emitted from fuel stored in attached garages. Overheated cooking oils emit acrolein and formaldehyde. A meta-analysis of 77 surveys of VOCs in homes in the US found the top ten riskiest indoor air VOCs were acrolein, formaldehyde, benzene, hexachlorobutadiene, acetaldehyde, 1,3-butadiene, benzyl chloride, 1,4-dichlorobenzene, carbon tetrachloride, acrylonitrile, and vinyl chloride. These compounds exceeded health standards in most homes.

Organic chemicals are widely used as ingredients in household products. Paints, varnishes, and wax all contain organic solvents, as do many cleaning, disinfecting, cosmetic, degreasing, and hobby products. Fuels are made up of organic chemicals. All of these products can release organic compounds during usage, and, to some degree, when they are stored. Testing emissions from building materials used indoors has become increasingly common for floor coverings, paints, and many other important indoor building materials and finishes.

Several initiatives envisage to reduce indoor air contamination by limiting VOC emissions from products. There are regulations in France and in Germany, and numerous voluntary ecolabels and rating systems containing low VOC emissions criteria such as EMICODE, M1, Blue Angel and Indoor Air Comfort in Europe, as well as California Standard CDPH Section 01350 and several others in the USA. These initiatives changed the marketplace where an increasing number of low-emitting products has become available during the last decades.

At least 18 Microbial VOCs (MVOCs) have been characterised including 1-octen-3-ol, 3-methylfuran, 2-pentanol, 2-hexanone, 2-heptanone, 3-octanone, 3-octanol, 2-octen-1-ol, 1-octene, 2-pentanone, 2-nonanone, borneol, geosmin, 1-butanol, 3-methyl-1-butanol, 3-methyl-2-butanol, and thujopsene. The first of these compounds is called mushroom alcohol. The last four are products of *Stachybotrys chartarum*, which has been linked with sick building syndrome.

Legionella

Legionellosis or Legionnaire's Disease is caused by a waterborne bacterium *Legionella* that grows best in slow-moving or still, warm water. The primary route of exposure is through the creation of an aerosol effect, most commonly from evaporative cooling towers or showerheads. A common source of Legionella in commercial buildings is from poorly placed or maintained evaporative cooling towers, which often release water in an aerosol which may enter nearby ventilation intakes. Outbreaks in medical facilities and nursing homes, where patients are immuno-suppressed and immuno-weak, are the most commonly reported cases of Legionellosis. More than one case has involved outdoor fountains in public attractions. The presence of Legionella in commercial building water supplies is highly under-reported, as healthy people require heavy exposure to acquire infection.

Legionella testing typically involves collecting water samples and surface swabs from evaporative cooling basins, shower heads, faucets/taps, and other locations where warm water collects. The samples are then cultured and colony forming units (cfu) of Legionella are quantified as cfu/Liter.

Legionella is a parasite of protozoans such as amoeba, and thus requires conditions suitable for both organisms. The bacterium forms a biofilm which is resistant to chemical and antimicrobial treatments, including chlorine. Remediation for Legionella outbreaks in commercial buildings vary, but often include very hot water flushes (160 °F; 70 °C), sterilisation of standing water in evaporative cooling basins, replacement of shower heads, and in some cases flushes of heavy metal salts. Preventative measures include adjusting normal hot water levels to allow for 120 °F (50 °C) at the tap, evaluating facility design layout, removing faucet aerators, and periodic testing in suspect areas.

Other Bacteria

There are many bacteria of health significance found in indoor air and on indoor surfaces. The role of microbes in the indoor environment is increasingly studied using modern gene-based analysis of environmental samples. Currently efforts are under way to link microbial ecologists and indoor air scientists to forge new methods for analysis and to better interpret the results.

"There are approximately ten times as many bacterial cells in the human flora as there are human cells in the body, with large numbers of bacteria on the skin and as gut flora." A large fraction of the bacteria found in indoor air and dust are shed from humans. Among the most important bacteria known to occur in indoor air are Mycobacterium tuberculosis, Staphylococcus aureus, Streptococcus pneumoniae.

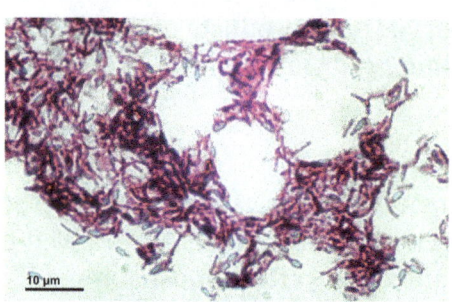

Bacteria (26 2 27) Airborne microbes

Asbestos Fibers

Many common building materials used before 1975 contain asbestos, such as some floor tiles, ceiling tiles, shingles, fireproofing, heating systems, pipe wrap, taping muds, mastics, and other insulation materials. Normally, significant releases of asbestos fiber do not occur unless the building materials are disturbed, such as by cutting, sanding, drilling, or building remodelling. Removal of asbestos-containing materials is not always optimal because the fibers can be spread into the air during the removal process. A management program for intact asbestos-containing materials is often recommended instead.

When asbestos-containing material is damaged or disintegrates, microscopic fibers are dispersed into the air. Inhalation of asbestos fibers over long exposure times is associated with increased incidence of lung cancer, in particular the specific form mesothelioma. The risk of lung cancer from inhaling asbestos fibers is significantly greater to smokers, however there is no confirmed connection to damage caused by asbestosis . The symptoms of the disease do not usually appear until about 20 to 30 years after the first exposure to asbestos.

Asbestos is found in older homes and buildings, but occurs most commonly in schools, hospitals and industrial settings. Although all asbestos is hazardous, products that are friable, eg. sprayed coatings and insulation, pose a significantly higher hazard as they are more likely to release fibers to the air. The US Federal Government and some states have set standards for acceptable levels of asbestos fibers in indoor air. There are particularly stringent regulations applicable to schools.

Carbon Dioxide

Carbon dioxide (CO_2) is a relatively easy to measure surrogate for indoor pollutants emitted by humans, and correlates with human metabolic activity. Carbon dioxide at levels that are unusually high indoors may cause occupants to grow drowsy, to get headaches, or to function at lower activity levels. Outdoor CO_2 levels are usually 350-450 ppm whereas the maximum indoor CO_2 level considered acceptable is 1000 ppm. Humans are the main indoor source of carbon dioxide in most buildings. Indoor

CO_2 levels are an indicator of the adequacy of outdoor air ventilation relative to indoor occupant density and metabolic activity.

To eliminate most complaints, the total indoor CO_2 level should be reduced to a difference of less than 600 ppm above outdoor levels. The National Institute for Occupational Safety and Health (NIOSH) considers that indoor air concentrations of carbon dioxide that exceed 1,000 ppm are a marker suggesting inadequate ventilation. The UK standards for schools say that carbon dioxide in all teaching and learning spaces, when measured at seated head height and averaged over the whole day should not exceed 1,500 ppm. The whole day refers to normal school hours (i.e. 9:00am to 3:30pm) and includes unoccupied periods such as lunch breaks. In Hong Kong, the EPD established indoor air quality objectives for office buildings and public places in which a carbon dioxide level below 1,000 ppm is considered to be good. European standards limit carbon dioxide to 3,500 ppm. OSHA limits carbon dioxide concentration in the workplace to 5,000 ppm for prolonged periods, and 35,000 ppm for 15 minutes. These higher limits are concerned with avoiding loss of consciousness (fainting), and do not address impaired cognitive performance and energy, which begin to occur at lower concentrations of carbon dioxide.

Carbon dioxide concentrations increase as a result of human occupancy, but lag in time behind cumulative occupancy and intake of fresh air. The lower the air exchange rate, the slower the buildup of carbon dioxide to quasi "steady state" concentrations on which the NIOSH and UK guidance are based. Therefore, measurements of carbon dioxide for purposes of assessing the adequacy of ventilation need to be made after an extended period of steady occupancy and ventilation - in schools at least 2 hours, and in offices at least 3 hours - for concentrations to be a reasonable indicator of ventilation adequacy. Portable instruments used to measure carbon dioxide should be calibrated frequently, and outdoor measurements used for calculations should be made close in time to indoor measurements. Corrections for temperature effects on measurements made outdoors may also be necessary.

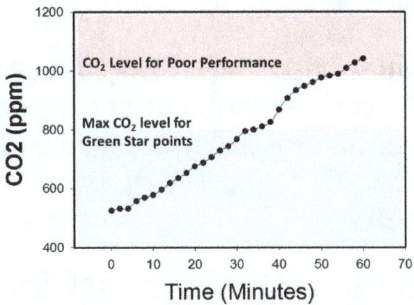

CO_2 levels in an enclosed office room can increase to over 1,000 ppm within 45 minutes.

Carbon dioxide concentrations in closed or confined rooms can increase to 1,000 ppm within 45 minutes of enclosure. For example, in a 3.5-by-4-metre (11 ft × 13 ft) sized office, atmospheric carbon dioxide increased from 500 ppm to over 1,000 ppm within 45 minutes of ventilation cessation and closure of windows and doors.

Ozone

Ozone is produced by ultraviolet light from the Sun hitting the Earth's atmosphere (especially in the ozone layer), lightning, certain high-voltage electric devices (such as air ionizers), and as a by-product of other types of pollution.

Ozone exists in greater concentrations at altitudes commonly flown by passenger jets. Reactions between ozone and onboard substances, including skin oils and cosmetics, can produce toxic chemicals as by-products. Ozone itself is also irritating to lung tissue and harmful to human health. Larger jets have ozone filters to reduce the cabin concentration to safer and more comfortable levels.

Outdoor air used for ventilation may have sufficient ozone to react with common indoor pollutants as well as skin oils and other common indoor air chemicals or surfaces. Particular concern is warranted when using "green" cleaning products based on citrus or terpene extracts, because these chemicals react very quickly with ozone to form toxic and irritating chemicals as well as fine and ultrafine particles. Ventilation with outdoor air containing elevated ozone concentrations may complicate remediation attempts.

Ozone is on the list of six criteria air pollutant list. The Clean Air Act of 1990 required the United States Environmental Protection Agency to set National Ambient Air Quality Standards (NAAQS) for six common indoor air pollutants harmful to human health. There are also multiple other organizations that have put forth air standards such as Occupational Safety and Health Administration (OSHA), National Institute for Occupational Safety and Health (NIOSH), and the World Health Organization (WHO). The OSHA standard for Ozone concentration within a space is 0.1 ppm . While the NAAQS and the EPA standard for ozone concentration is limited to 0.07 ppm. . The type of ozone being regulated is ground-level ozone that is within the breathing range of most building occupants.

Particulates

Atmospheric particulate matter, also known as particulates, can be found indoors and can affect the health of occupants. Authorities have established standards for the maximum concentration of particulates to ensure indoor air quality.

Prompt Cognitive Deficits

In 2015, experimental studies reported the detection of significant episodic (situational) cognitive impairment from impurities in the air breathed by test subjects who were not informed about changes in the air quality. Researchers at the Harvard University and SUNY Upstate Medical University and Syracuse University measured the cognitive performance of 24 participants in three different controlled laboratory atmospheres that simulated those found in "conventional" and "green" buildings, as well as green buildings with enhanced ventilation. Performance was evaluated objectively using the widely used Strategic Management Simulation software simulation tool, which is a well-validated assessment test for executive decision-making in an unconstrained situation allowing initiative and improvisation. Significant deficits were observed in the performance scores achieved in increasing concentrations of either volatile organic compounds (VOCs) or carbon dioxide, while keeping other factors constant. The highest impurity levels reached are not uncommon in some classroom or office environments.

Effect of Indoor Plants

Spider plants *(Chlorophytum comosum)* absorb some airborne contaminants

Houseplants together with the medium in which they are grown can reduce components of indoor air pollution, particularly volatile organic compounds (VOC) such as benzene, toluene, and xylene. Plants remove CO_2 and release oxygen and water, although the quantitative impact for house plants is small. Most of the effect is attributed to the growing medium alone, but even this effect has finite limits associated with the type and quantity of medium and the flow of air through the medium. The effect of house plants on VOC concentrations was investigated in one study, done in a static chamber, by NASA for possible use in space colonies. The results showed that the removal of the challenge chemicals was roughly equivalent to that provided by the ventilation that occurred in a very energy efficient dwelling with a very low ventilation rate, an air exchange rate of about 1/10 per hour. Therefore, air leakage in most homes, and in non-residential buildings too, will generally remove the chemicals faster than the researchers reported for the plants tested by NASA. The most effective household plants reportedly included aloe vera, English ivy, and Boston fern for removing chemicals and biological compounds.

Plants also appear to reduce airborne microbes and molds, and to increase humidity. However, the increased humidity can itself lead to increased levels of mold and even VOCs.

When carbon dioxide concentrations are elevated indoors relative to outdoor concentrations, it is only an indicator that ventilation is inadequate to remove metabolic products associated with human occupancy. Plants require carbon dioxide to grow and release oxygen when they consume carbon dioxide. A study considered uptake rates of ketones and aldehydes by the peace lily (*Spathiphyllum clevelandii*) and golden pothos (*Epipremnum aureum*) Akira Tani and C. Nicholas Hewitt found "Longer-term fumigation results revealed that the total uptake amounts were 30–100 times as much as the amounts dissolved in the leaf, suggesting that volatile organic carbons are metabolized in the leaf and/or translocated through the petiole." It is worth noting the researchers sealed the plants in Teflon bags. "No VOC loss was detected from the bag when the plants were absent. However, when the plants were in the bag, the levels of aldehydes and ketones both decreased slowly but continuously, indicating removal by the plants". Studies done in sealed bags do not faithfully reproduce the conditions in the indoor environments of interest. Dynamic conditions with outdoor air ventilation and the processes related to the surfaces of the building itself and its contents as well as the occupants need to be studied.

While results do indicate house plants may be effective at removing some VOCs from air supplies, a review of studies between 1989 and 2006 on the performance of house-plants as air cleaners, presented at the Healthy Buildings 2009 conference in Syracuse, New York, concluded "indoor plants have little, if any, benefit for removing indoor air of VOC in residential and commercial buildings."

Since high humidity is associated with increased mold growth, allergic responses, and respiratory responses, the presence of additional moisture from houseplants may not be desirable in all indoor settings.

Types of Symptoms and Complaints

The effects of IAQ problems are often nonspecific symptoms rather than clearly defined illnesses. Symptoms commonly attributed to IAQ problems include:

- headache
- fatigue
- shortness of breath
- sinus congestion
- cough
- sneezing
- eye, nose, and throat irritation
- skin irritation
- dizziness
- nausea

All of these symptoms, however, may also be caused by other factors, and are not necessarily due to air quality deficiencies.

"Health" and "comfort" are used to describe a spectrum of physical sensations. For example, when the air in a room is slightly too warm for a person's activity level, that person may experience mild discomfort. If the temperature continues to rise, discomfort increases and symptoms such as fatigue, stuffiness, and headaches can appear.

Some complaints by building occupants are clearly related to the discomfort end of the spectrum. One of the most common IAQ complaints is that "there's a funny smell in here." Odors are often associated with a perception of poor air quality, whether or not they cause symptoms. Environmental stressors such as improper lighting, noise, vibration, overcrowding, ergonomic stressors, and job-related psychosocial problems (such as job stress) can produce symptoms that are similar to those associated with poor air quality.

The term sick building syndrome (SBS) is sometimes used to describe cases in which building occupants experience acute health and comfort effects that are apparently linked to the time they spend in the building, but in which no specific illness or cause can be identified. The complaints may be localized in a particular room or zone or may be widespread throughout the building. Many different symptoms have been associated with SBS, including respiratory complaints, irritation, and fatigue. Analysis of air samples often fails to detect high concentrations of specific contaminants. The problem may be caused by any or all of the following:

- the combined effects of multiple pollutants at low concentrations

- other environmental stressors

  (e.g., overheating, poor lighting, noise)

- ergonomic stressors

- job-related psychosocial stressors

  (e.g., overcrowding, labor-managementproblems)

- unknown factors

Building-related illness (BRI) is a term referring to illness brought on by exposure to the building air, where symptoms of diagnosable illness are identified (e.g., certain allergies or infections) and can be directly attributed to environmental agents in the air. Legionnaire's disease and hypersensitivity pneumonitis are examples of BRI that can have serious, even lifethreatening consequences.

A small percentage of the population may be sensitive to a number of chemicals in indoor air, each of which may occur at very low concentrations. The existence of this condition, which is known as multiple chemical sensitivity (MCS), is a matter of considerable controversy. MCS is not currently recognized by the major medical organizations, but medical opinion is divided. The applicability of access for the disabled and worker's compensation regulations to people who believe they are chemically sensitive may become concerns for facility managers

Sometimes several building occupants experience rare or serious health problems (e.g., cancer, miscarriages, Lou Gehrig's disease) over a relatively short time period. These clusters of health problems are occasionally blamed on indoor air quality, and can produce tremendous anxiety among building occupants. State or local Health Departments can provide advice and assistance if clusters are suspected. They may be able to help answer key questions such as whether the apparent cluster is actually unusual and whether the underlying cause could be related to IAQ.

Ventilation

Ventilation is the process by which 'clean' air (normally outdoor air) is intentionally provided to a space and stale air is removed. This may be accomplished by either natural or mechanical means.

Air infiltration and exfiltration: In addition to intentional ventilation, air inevitably enters a building by the process of 'air infiltration'. This is the uncontrolled flow of air into a space through adventitious or unintentional gaps and cracks in the building envelope. The corresponding loss of air from an enclosed space is termed 'exfiltration'. The rate of air infiltration is dependent on the porosity of the building shell and the magnitude of the natural driving forces of wind and temperature. Vents and other openings incorporated into a building as part of ventilation design can also become routes for unintentional air flow when the pressures acting across such openings are dominated by weather conditions rather than intentionally (e.g. mechanically) induced driving forces. Air infiltration not only adds to the quantity of air entering the building but may also distort the intended air flow pattern to the detriment of overall indoor air quality and comfort. Although the magnitude of air infiltration can be considerable, it is frequently ignored by the designer. The consequences are inferior performance, excessive energy consumption, an inability to provide adequate heating (or cooling) and drastically impaired performance from heat recovery devices. Some Countries have introduced air-tightness Standards to limit infiltration losses (Limb 1994).

Other air losses, e.g. duct leakage: Air leakage from the seams and joints of ventilation, heating and air conditioning circulation ducts can be substantial. When, as is common, such ducting passes through unconditioned spaces, significant energy loss may occur. Modera (1993), for example, estimates that as much as 20% of the heat from typical North American domestic warm air heating systems can be lost through duct leakage. Pollutants may also be drawn into the building through these openings. As a consequence, considerable research and development into the performance of duct sealing measures is being undertaken.

Air recirculation: Air recirculation is frequently used in commercial buildings to provide for thermal conditioning. Recirculated air is usually filtered for dust removal but, since oxygen is not replenished and metabolic pollutants are not removed, recirculation should not usually be considered as contributing towards ventilation need.

Ventilation is needed to provide oxygen for metabolism and to dilute metabolic pollutants (carbon dioxide and odour). It is also used to assist in maintaining good indoor air quality by diluting and removing other pollutants emitted within a space but should not be used as a substitute for proper source control of pollutants. Ventilation is additionally used for cooling and (particularly in dwellings) to provide oxygen to combustion appliances. Good ventilation is a major contributor to the health and comfort of building occupants.

To guarantee a good indoor air quality it is necessary to remove the "consumed air" (exhaust air) and to supply "fresh air" (supply air). The different kinds of air infiltration are shown in the following figure.

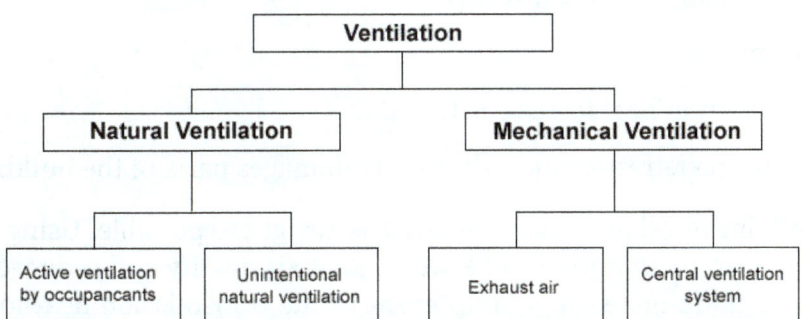

The possible kinds of ventilation are versatile . At any rate the concept of the ventilation should be adjusted to the remaining building services

Pressure differences are the physical reason of each flow and therewith of the air exchange. They can be produced by fans for mechanical ventilation or by natural driving forces for "natural" ventilation.

Natural Ventilation

Natural ventilation has to be distinguished from intended natural ventilation and unintentional natural ventilation. In both cases the driving forces for air exchange are caused by differences in temperature and the wind field. Intended natural ventilation results from:

- *Stack ventilation:* The existing buoyancy inside a gap is used to lead air away *(stack effect)* and to supply the room with fresh air inlets at the same time.

- *Window ventilation:* "shock ventilation" means a short and intensive ventilation through totally open windows; "permanent ventilation" means a long lasting ventilation by tilted windows.

Especially shock ventilation is very effective by opening two opposite windows *(cross ventilation)*. The frequency and duration of ventilation through wind strongly depend on the personal feeling or attitude of the occupants. Therefore, it is very difficult to realise a demand oriented and optimised ventilation through windows because there is either

- too much ventilation, which causes high heat losses or

- too little ventilation, which leads to problems of air quality and damages resulting from humidity.

Unintended natural ventilation (infiltration / exfiltration) results from leakages in the building envelope (joints, leaky windows and doors, installation). In many old buildings the unintended natural ventilation is almost sufficient for the supply with fresh air.

As a consequence, the following disadvantages arise:

- this air exchange is uncontrollable,

- drafts can arise,

- high ventilation heat losses occur and

- convective penetration due to humidity damages parts of the building.

Today, the building envelope is implemented as airtight as possible. Using a controlled ventilation will lead to an optimum between good air quality and reduced ventilation heat losses, which are increasingly important as the transmission heat losses are becoming smaller due to the meanwhile high thermal insulation standard.

Mechanical Ventilation

Fans produce the pressure difference, which is necessary for the air exchange. This makes it possible to control the air exchange and guarantee a hygienically indoor air quality. Mechanical ventilation units are installed in different versions:

Central exhaust ventilation systems are operated with one fan only, which produces a negative pressure in the ventilated section. The supply air flows from outside through leakages or holes into the ventilated section. The exhaust air is blown outside. In order to utilise the heat of the exhaust air, it can be led to heat pumps in bigger systems.

Central ventilation systems have two fans. One fan supplies the building with just the exact quantity of fresh air which the other fan extracts. In buildings, the air inlet and the exhaust openings are not necessarily installed in the same room. Insignificant differences in pressure between supply air (living room, dining room, etc.) and exhaust air section (kitchen, bathroom, etc.) transfer air from the supply zone into the exhaust zone. The warm exhaust air can be used to heat the supply air when the supply air and exhaust air volume rate is directed to one another through a heat exchanger. This recovery is called *recuperative heat recovery*. The different heat exchanger can be distinguished from their air ducts as parallel flow, reverse flow and cross flow heat exchanger. The different types can also be combined, e.g. to a cross reverse flow heat exchanger.

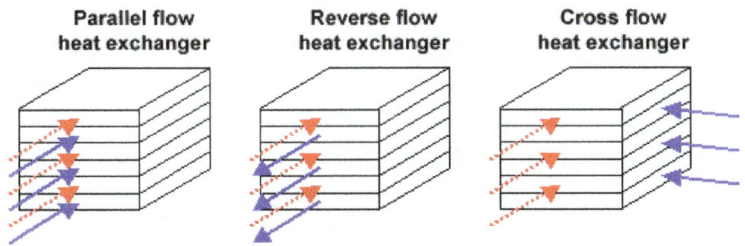

Schematical demonstration of the different air ducts of heat exchangers.

Different Characterisation of Air

*   outdoor inlet air is called fresh air,

*   air delivered to indoors is called supply air,

*   sucked off indoor air is called extract air and

*   air delivered outdoors is called exhaust air (Figure 3).

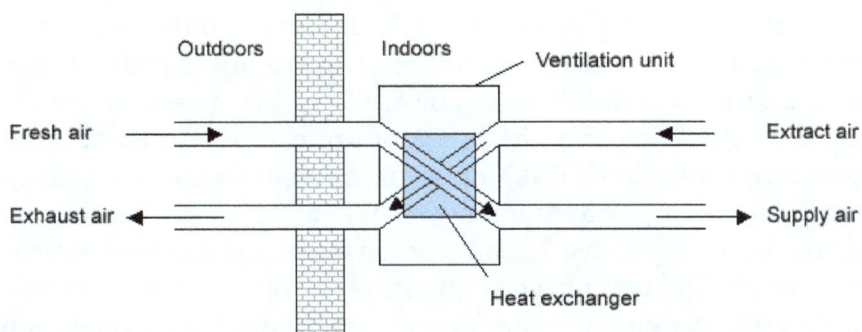

The cold outdoor air is blown through the heat exchanger, warmed up by the exhaust air and injected as supply air into the ventilated room. The cooled off extract air is blown outside as exhaust air

Regenerative heat recovery uses the principle of heat storage. The waste heat of the extract air is stored in a buffer and transferred to the supply air later. Central ventilation units of good quality can reach a heat recovery efficiency of 90 %. Today, these central ventilation systems offer the best facilities for supplying a building with fresh air and reducing ventilation heat losses at the same time.

A clever planned duct system, which is required for the already described *central ventilation systems*, is not necessary for *decentralised ventilation units*. Therefore, they are suitable for retrofit of buildings or specific ventilation of a single room in which a high pollutant concentration or air humidity can occur. The characterised functions of exhaust air ventilation systems or central ventilation units can also be realised with decentralised ventilation units. Of course decentralised ventilation units do not attain the quite high heat recovery efficiency of central systems. In addition, unsuited exhaust and supply openings can cause short-circuit flow through unfavourable installation.

Normally, the use of mechanical ventilation systems with heat recovery is only reasonable if the building envelope is airtight, so that uncontrolled infiltration, caused by leakage, is very small. If the supply and exhaust airflow rates are badly balanced, the ventilation system puts the building under a pressure difference against outdoor conditions. Then, air may be leaking through the building envelope and increase the ventilation heat losses.

Additional Options

A ventilation system can take over additional tasks. Especially in passive houses, halls and industrial buildings, the existing ventilation system is used for heating *(air-heating)*. In passive houses the outside air, warmed up by passing the heat recovery system, is blown through a *"re-heating system"* and distributed in the ventilated space. The heating power can be calculated by the following equation:

$$H = \rho \cdot c \cdot Q \cdot (T_{air} - T_{space})$$

One can see, that the heating power H can be increased only by an enhancement of the airflow rate Q or the air temperature. The specific density ρ the heat capacity of air c and the inside air temperature T_{space} are given. If the airflow rate depends on the ventilation, only the temperature can be increased. The increase of the temperature is restricted due to reasons of comfort and air quality, because the carbonization of dust particles already occurs at 52°C. Air heating systems are operated particularly in passive houses, because the attainable heating power is comparatively small. The two physical quantities Q and T_{air} can be varied in larger extend, if air-heating systems are partly used in recirculating air operations in halls and industrial plants.

Air conditioning is attained by upgrading the air heating system with an air cooler and a moisture regulator. Thus, all air treatments like ventilating, cooling, heating, humidifying and dehumidifying can be performed. Such systems are used both as comfort units for living and work space (production of a convenient indoor climate), and, even more often, as industrial air conditioning (observance of air parameters to produce or to lodge a product).

Earth Heat Exchanger

An earth heat exchanger directs the outdoor air through a pipe which is installed in the ground. Thereby, a heat exchange of the air with the soil through the pipe wall takes place. The pre-heated or pre-cooled air is led into the ventilation system of the building.

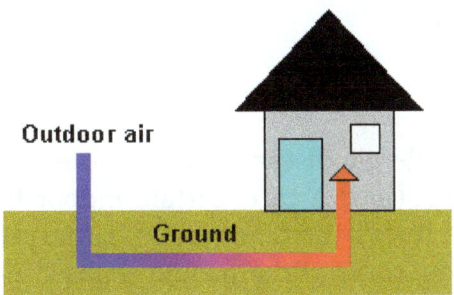

Scheme of an earth heat exchanger.

In winter, the cold outdoor air is pre-heated through the ground and the heating load is reduced. In summer, the warm outdoor air is cooled by the earth heat exchanger, which reduces the cooling load of the building.

Night Ventilation

Night ventilation as a concept can be installed in order to reduce, or even to avoid the overheating in summer. Therefore, either natural ventilation or the mechanical ventilation can be implemented. At night, the building is ventilated with cool outdoor air in order to reduce the indoor air temperature.

Indoor Bioaerosol

Bioaerosols are microorganisms or particles, gases, vapors, or fragments of biological origin (i.e., alive or released from a living organism) that are in the air. Bioaerosols are everywhere in the environment

In order for microorganisms to release indoor bioaerosols, they must get indoors, grow and multiply on some material and then get into the air. Microorganisms can get indoors through the heating, ventilation, and air conditioning system, doors, windows, cracks in the walls, the potable drinking water system, or be brought in on the shoes and clothes of people working or visiting in the building. Water, humidity, temperature, nutrient sources (e.g., sheetrock, wood paneling, cellulose ceiling tiles, carpets, upholstered furniture, and fiberglass-lined air ducts) and oxygen determine whether micro-organisms will grow in the indoor environment. The most common microorganisms found indoors are fungi and bacteria.

The Diseases Caused by Indoor Bioaerosols

Bioaerosols enter the human body mostly through being breathed in. So, the diseases they cause usually affect the respiratory system.

The diseases caused by indoor bioaerosols fall into two categories: hypersensitivity diseases and infectious diseases.

Hypersensitivity Diseases

Hypersensitivity diseases (allergic diseases) result from exposure to materials in the environment called antigens (in this case, certain indoor bioaerosols) that stimulate an allergic response by the body's immune system. Some people are more susceptible than others. In other words, some of the people exposed may become ill and others may not. These diseases usually are diagnosed by a physician. Once an individual has developed a hypersensitivity disease, a very small amount of the antigen may cause a severe reaction. Hypersensitivity diseases account for most of the health problems due to indoor bioaerosols.

- Building-related asthma may result in complaints of chest tightness, wheezing, coughing, and shortness of breath. These symptoms may occur within an hour of exposure or 4-12 hours after exposure. Building-related asthma can be caused by airborne fungi such as Alternaria, glycoproteins from fungi, proteases (digestive enzymes that cause the breakdown of proteins) from bacteria, the algae Clorococus, ragweed pollen, dust mites, and dander from cats.

- Allergic rhinitis involves stuffiness of the nose, clear discharge from the nose, itchy nose, and sneezing. Itching and puffy eyes may also occur. All the indoor bioaerosols listed under buildingrelated asthma except the bacteria proteases also cause rhinitis.

- Hypersensitivity pneumonitis (extrinsic allergic alveolitis) can be an acute, recurrent pneumonia with fever, cough, chest tightness, and fluids entering the lungs. Or, it can be a cough that progresses to shortness of breath, fatigue, weight loss and thickening and scarring of the lungs. The microorganisms associated with hypersensitivity pneumonitis are fungi such as Penicillium and Sporobolomyces,bacteria such as Thermoactinomyces, and protozoa such as Acanthamoeba.

- Humidifier fever results in fever, chills, muscle aches, and malaise (general feeling of being unwell), but no lung symptoms. The symptoms usually start within 4-8 hours of exposure and end within 24 hours without long-term effects.

Infectious Diseases

Infectious diseases are caused by the invasion of the body by a harmful organism. Some examples of infectious diseases caused by indoor bioaerosols follow.

- Legionnaire's disease, a bacterial pneumonia, is caused by Legionellapneu-mophila. It is a type of pneumonia that affects the lungs and may also affect the stomach and intestines, kidneys, and central nervous system. It can take 2-10 days after exposure to develop and frequently requires hospitalization. The source of the disease has been traced to aerosols from contaminated cooling towers, evaporative condensers, whirlpools, shower heads, faucets, and hot water tanks.

- Pontiac fever is also caused by Legionella. Pontiac fever is a "flu-like" illness with fever, chills, headache, myalgia (pain in the muscles), cough, nausea, and breathlessness. Pneumonia does not occur. It usually lasts 2-5 days. The sources are the same as for Legionnaire's disease.

- Histoplasmosis and Cryptococcosis, both fungal infections, may occur when contaminated bird droppings enter the indoor environment. Infection with Histoplasma often results in no symptoms or there may be mild respiratory illness (cough, fever, malaise). Rarely, a life threatening illness involving many parts of the body occurs. Infection with Cryptococcus results in inflammation of the brain and the membranes covering it and also can involve the lungs, kidneys, prostate gland, bones, or liver. The skin may also be affected with acne-like lesions, ulcers, or tumorlike masses.

Passive Smoking

Passive smoking means breathing in other people's tobacco smoke. Exhaled smoke is called exhaled mainstream smoke. The smoke drifting from a lit cigarette is called sidestream smoke. The combination of mainstream and sidestream smoke is called second-hand smoke (SHS) or environmental tobacco smoke (ETS). Second-hand smoke is a serious health risk for both those who smoke and those who do not. Children are particularly at risk of serious health effects from second-hand smoke.

There are three types of cigarette smoke:

1. The smoke that the smoker inhales directly.

2. The secondary smoke that is emitted sideways from the cigarette (and which has a very different composition from the smoke directly inhaled by the smoker).

3. The smoke exhaled by the smoker, which plays a minor role.

The secondary smoke is the most direct cause of passive smoking. It is emitted over a long period of time (about 10 minutes, while the smoke exhaled by the smoker is only emitted for between 20 and 30 seconds) and most importantly, it has not been filtered by either the cigarette or the lungs of the smoker.

The smoker inhales more smoke than the passive smoker. However, two further factors must be taken into consideration:

1. The composition of the smoke. When it is emitted, the secondary smoke contains a greater number of harmful substances than the smoke inhaled by the active smoker. The concentration of substances drops dramatically in the outside air. The passive smoker therefore inhales a stream of gases in which chemical products are present in varying quantities.

2. The length of exposure. This may be occasional (eg. the length of an evening from time to time) or regular (from childhood through to adulthood). Obviously, the longer the person is exposed to cigarette smoke, the more health risks.

Pathophysiology

A 2004 study by the International Agency for Research on Cancer of the World Health Organization concluded that non-smokers are exposed to the same carcinogens as active smokers. Sidestream smoke contains more than 4,000 chemicals, including 69 known carcinogens. Of special concern are polynuclear aromatic hydrocarbons, tobacco-specific N-nitrosamines, and aromatic amines, such as 4-aminobiphenyl, all known to be highly carcinogenic. Mainstream smoke, sidestream smoke, and second-hand smoke contain largely the same components, however the concentration varies depending on type of smoke. Several well-established carcinogens have been shown by the tobacco companies' own research to be present at higher concentrations in sidestream smoke than in mainstream smoke.

Second-hand smoke has been shown to produce more particulate-matter (PM) pollution than an idling low-emission diesel engine. In an experiment conducted by the Italian National Cancer Institute, three cigarettes were left smoldering, one after the other, in a 60 m³ garage with a limited air exchange. The cigarettes produced PM pollution exceeding outdoor limits, as well as PM concentrations up to 10-fold that of the idling engine.

Second-hand tobacco smoke exposure has immediate and substantial effects on blood and blood vessels in a way that increases the risk of a heart attack, particularly in people already at risk. Exposure to tobacco smoke for 30 minutes significantly reduces coronary flow velocity reserve in healthy nonsmokers. Second-hand smoke is also associated with impaired vasodilation among adult nonsmokers. Second-hand smoke exposure also affects platelet function, vascular endothelium, and myocardial exercise tolerance at levels commonly found in the workplace.

Pulmonary emphysema can be induced in rats through acute exposure to sidestream tobacco smoke (30 cigarettes per day) over a period of 45 days. Degranulation of mast cells contributing to lung damage has also been observed.

The term "third-hand smoke" was recently coined to identify the residual tobacco smoke contamination that remains after the cigarette is extinguished and second-hand smoke has cleared from the air. Preliminary research suggests that by-products of third-hand smoke may pose a health risk, though the magnitude of risk, if any, remains unknown. In October 2011, it was reported that Christus St. Frances Cabrini Hospital in Alexandria, Louisiana would seek to eliminate third-hand smoke beginning in July 2012, and that employees whose clothing smelled of smoke would not be allowed to work. This prohibition was enacted because third-hand smoke poses a special danger for the developing brains of infants and small children.

In 2008, there were more than 161,000 deaths attributed to lung cancer in the United States. Of these deaths, an estimated 10% to 15% were caused by factors other than first-hand smoking; equivalent to 16,000 to 24,000 deaths annually. Slightly more than half of the lung cancer deaths caused by factors other than first-hand smoking were found in nonsmokers. Lung cancer in non-smokers may well be considered one of the most common cancer mortalities in the United States. Clinical epidemiology of lung cancer has linked the primary factors closely tied to lung cancer in non-smokers as exposure to second-hand tobacco smoke, carcinogens including radon, and other indoor air pollutants.

Smoke-free Laws

As a consequence of the health risks associated with second-hand smoke, smoke-free regulations in indoor public places, including restaurants, cafés, and nightclubshave been introduced in a number of jurisdictions, at national or local level, as well as some outdoor open areas. Ireland was the first country in the world to institute a comprehensive national smoke-free law on smoking in all indoor workplaces on 29 March 2004. Since then, many others have followed suit. The countries which have ratified the WHO Framework Convention on Tobacco Control (FCTC) have a legal obligation to implement *effective* legislation "for protection from exposure to tobacco smoke in indoor workplaces, public transport, indoor public places and, as appropriate, other public places". (Article 8 of the FCTC) The parties to the FCTC have further adopted *Guidelines on the Protection from Exposure to Second-hand Smoke* which state that "effective measures to provide protection from exposure to tobacco smoke require the total elimination of smoking and tobacco smoke in a particular space or environment in order to create a 100% smoke-free environment".

Opinion polls have shown considerable support for smoke-free laws. In June 2007, a survey of 15 countries found 80% approval for smoke-free laws. A survey in France, reputedly a nation of smokers, showed 70% support.

Effects

Smoking bans by governments result in decreased harm from second hand smoke including decrease cardiovascular disease. In the first 18 months after the town of Pueblo,

Colorado enacted a smoke-free law in 2003, hospital admissions for heart attacks dropped 27%. Admissions in neighbouring towns without smoke-free laws showed no change, and the decline in heart attacks in Pueblo was attributed to the resulting reduction in second-hand smoke exposure. A 2004 smoking ban instituted in Massachusetts workplaces decreased workers' secondhand smoke exposure from 8% of workers in 2003 to 5.4% of workers in 2010. A 2016 review also found benefits of decrease exposure to smoke from specific location policies.

In 2001, a systematic review for the Guide to Community Preventative Services acknowledged strong evidence of the effectiveness of smoke-free policies and restrictions in reducing expose to second-hand smoke. A follow up to this review, identified the evidence on which the effectiveness of smoking bans reduced the prevalence of tobacco use. The examined studies provided sufficient evidence that smoke-free policies reduce tobacco use among workers when implemented in worksites or by communities.

While a number of studies funded by the tobacco industry have claimed a negative economic impact from smoke-free laws, no independently funded research has shown any such impact. A 2003 review reported that independently funded, methodologically sound research consistently found either no economic impact or a positive impact from smoke-free laws.

Air nicotine levels were measured in Guatemalan bars and restaurants before and after an implemented smoke-free law in 2009. Nicotine concentrations significantly decreased in both the bars and restaurants measured. Also, the employees support for a smoke-free workplace substantially increased in the post-implementation survey compared to pre-implementation survey. The result of this smoke-free law provides a considerably more healthy work environment for the staff.

Public Opinion

Recent surveys taken by the Society for Research on Nicotine and Tobacco demonstrates supportive attitudes of the public, towards smoke-free policies in outdoor areas. A vast majority of the public supports restricting smoking in various outdoor settings. The respondents reasons for supporting the policies were for varying reasons such as, litter control, establishing positive smoke-free role models for youth, reducing youth opportunities to smoke, and avoiding exposure to secondhand smoke.

Alternative Forms

Alternatives to smoke-free laws have also been proposed as a means of harm reduction, particularly in bars and restaurants. For example, critics of smoke-free laws cite studies suggesting ventilation as a means of reducing tobacco smoke pollutants and improving air quality. Ventilation has also been heavily promoted by the tobacco industry as an alternative to outright bans, via a network of ostensibly independent experts with often undisclosed ties to the industry. However, not all critics have connections to the industry.

The American Society of Heating, Refrigerating and Air-Conditioning Engineers (ASHRAE) officially concluded in 2005 that while completely isolated smoking rooms do eliminate the risk to nearby non-smoking areas, smoking bans are the only means of completely eliminating health risks associated with indoor exposure. They further concluded that no system of dilution or cleaning was effective at eliminating risk. The U.S. Surgeon General and the European Commission Joint Research Centrehave reached similar conclusions. The implementation guidelines for the WHO Framework Convention on Tobacco Control states that engineering approaches, such as ventilation, are ineffective and do not protect against second-hand smoke exposure. However, this does *not* necessarily mean that such measures are useless in reducing harm, only that they fall short of the goal of reducing exposure completely to zero.

Others have suggested a system of tradable smoking pollution permits, similar to the cap-and-trade pollution permits systems used by the Environmental Protection Agency in recent decades to curb other types of pollution. This would guarantee that a portion of bars/restaurants in a jurisdiction will be smoke-free, while leaving the decision to the market.

Irritant Effects of Passive Smoking

Tobacco smoke inside a room tends to hang in mid-air rather than disperse. Hot smoke rises, but tobacco smoke cools rapidly, which stops its upward climb. Since the smoke is heavier than the air, the smoke starts to descend.

A person who smokes heavily indoors creates a low-lying smoke cloud that other house-holders have no choice but to breathe.

Tobacco smoke contains around 7,000 chemicals, made up of particles and gases, over 50 of which are known to cause cancer. Second-hand smoke has been confirmed as a cause of lung cancer in humans by several leading health authorities.

Compounds such as ammonia, sulphur and formaldehyde irritate the eyes, nose, throat and lungs. These compounds are especially harmful to people with respiratory conditions such as bronchitis or asthma. Exposure to second-hand smoke can either trigger or worsen symptoms.

Health Risks of Passive Smoking – Pregnant Women and Unborn Babies

Australian data indicates that about 11 per cent of women smoke during pregnancy. Both smoking and passive smoking can seriously affect the developing fetus.

Health risks for mothers who smoke during pregnancy include increased risk of:

- miscarriage and stillbirth;
- ectopic pregnancy;

- premature birth and low birth weight;

- sudden unexpected death in infants (SUDI), which includes sudden infant death syndrome (SIDS) and fatal sleep accidents;

- complications during birth.

A non-smoking pregnant woman is more likely to give birth earlier, and to a baby with a slightly lower birth weight if she is exposed to second-hand smoke in the home – for example, if her partner smokes.

Health Risks of Passive Smoking – Children

Children are especially vulnerable to the damaging effects of second-hand smoke. Some of the many health risks include:

- Passive smoking is a cause of sudden unexpected death in infants (SUDI), which includes sudden infant death syndrome (SIDS) and fatal sleep accidents.

- A child who lives in a smoking household for the first 18 months of their life has an increased risk of developing a range of respiratory illnesses, including bronchitis,bronchiolitis and pneumonia. They are also more prone to getting colds, coughs and glue ear (middle ear infections). Their lungs are weaker and do not grow to their full potential.

- A child exposed to second-hand smoke in the home is more likely to develop asthma symptoms, have more asthma attacks and use asthma medications more often and for a longer period.

- School-aged children of people who smoke are more likely to have symptoms such as cough, phlegm, wheeze and breathlessness.

- Children of people who smoke have an increased risk of meningococcal disease, which can sometimes cause death or disability.

Health Risks of Passive Smoking – Partners who have never Smoked

People who have never smoked who live with people who do smoke are at increased risk of a range of tobacco-related diseases and other health risks, including:

- Passive smoking increases the risk of heart disease. There is consistent evidence that people who do not smoke, who live in a smoky household, have higher risks of coronary heart disease than those who do not.

- Passive smoking makes the blood more 'sticky' and likely to clot, thereby leading to increased risk of various health conditions, including heart attack and stroke.

- There is evidence that passive smoking is associated with lower levels of antioxidant vitamins in the blood.

- Just 30 minutes of exposure to second-hand smoke can affect how your blood vessels regulate blood flow, to a similar degree to that seen in people who smoke.

- Long-term exposure to passive smoking may lead to the development of athero-sclerosis (narrowing of the arteries).

- People who do not smoke who suffer long-term exposure to second-hand smoke have a 20 to 30 per cent higher risk of developing lung cancer.

- There is increasing evidence that passive smoking can increase the risk of nasal sinus cancer, throat cancer, larynx cancer, breast cancer, long- and short-term respiratory symptoms, loss of lung function, and chronic obstructive pulmonary disease among people who do not smoke.

- It is estimated that in Australia, in the financial year 2004–05, 113 adults and 28 infants died from diseases caused by second-hand smoke in the home. Passive smoking – a good reason to quit.

The risks of active smoking are well known. If a person who smokes can't give up for their own health, then the health of their family or other members of their household could be a stronger motivation.

Reducing the Risk of Passive Smoking

If a person who smokes is unwilling or unable to stop immediately, there are various ways to help protect the health of the people with whom they live. Suggestions include:

- Make your home smoke-free. Limiting your smoking to one or two rooms is not an effective measure – tobacco smoke can easily drift through the rest of the house.

- Make sure that visitors to your house smoke their cigarettes outdoors.

- Make your car smoke-free. The other occupants will still be exposed to tobacco smoke even if the windows are open. In Victoria, it is illegal to smoke in cars carrying people who are under 18 years of age.

- Don't allow smoking in any enclosed space where people who do not smoke spend time – for example, in the garage, shed, cubby house, boat or caravan.

- Try to avoid taking children to outdoor areas where people are smoking and you can't easily move away, such as a café courtyard.

- Make sure that all people who look after your children provide a smoke-free environment.

References

- May, Jeffrey C. (2006). My office is killing me! : the sick building survival guide. Baltimore: The Johns Hopkins University Press. ISBN 978-0-8018-8342-2

- Samet JM (2008). "Secondhand smoke: facts and lies". Salud Pública De México. 50 (5): 428–34. doi:10.1590/S0036-36342008000500016. PMID 18852940

- "Proposed Identification of Environmental Tobacco Smoke as a Toxic Air Contaminant". California Environmental Protection Agency. 2005-06-24. Retrieved 2009-01-12

- Diethelm, PA; McKee, M (2009). "Denialism: what is it and how should scientists respond?". European Journal of Public Health. 19 (1): 2–4. doi:10.1093/eurpub/ckn139. PMID 19158101. Lay summary

- Tichenor, B. (1996). Characterizing Sources of Indoor Air Pollution and Related Sink Effects. ASTM STP 1287. West Conshohocken, PA: ASTM. ISBN 0-8031-2030-3

- McClure JB (April 2002). "Are biomarkers useful treatment aids for promoting health behavior change? An empirical review". Am J Prev Med. 22 (3): 200–7. doi:10.1016/S0749-3797(01)00425-1. PMID 11897465

- Rabin, Roni Caryn (2009-01-02). "A New Cigarette Hazard: 'Third-Hand Smoke'". New York Times. Retrieved 2009-01-12

- Bentayeb, M; Simoni, M; Norback, D; Baldacci, S; Maio, S; Viegi, G; Annesi-Maesano, I (2013). "Indoor air pollution and respiratory health in the elderly". Journal of Environmental Science and Health, Part A. 48 (14): 1783–9. doi:10.1080/10934529.2013.826052. PMID 24007433

Chapter 4

Air Quality Index

An important index for communicating the actual or predicted measure of air pollution is the air quality index. It is obtained by computing the air pollutant concentration over a specified averaging period. This chapter discusses in detail the different ways to measure air quality, such as air quality index, pollutant standards index, NowCast, Beta attenuation monitoring, etc.

The AQI is an index for reporting daily air quality. It tells you how clean or polluted your air is, and what associated health effects might be a concern for you. The AQI focuses on health effects you may experience within a few hours or days after breathing polluted air. EPA calculates the AQI for five major air pollutants regulated by the Clean Air Act: ground-level ozone, particle pollution (also known as particulate matter), carbon monoxide, sulfur dioxide, and nitrogen dioxide. For each of these pollutants, EPA has established national air quality standards to protect public health. Ground-level ozone and airborne particles are the two pollutants that pose the greatest threat to human health in this country.

Working of AQI

Think of the AQI as a yardstick that runs from 0 to 500. The higher the AQI value, the greater the level of air pollution and the greater the health concern. For example, an AQI value of 50 represents good air quality with little potential to affect public health, while an AQI value over 300 represents hazardous air quality.

An AQI value of 100 generally corresponds to the national air quality standard for the pollutant, which is the level EPA has set to protect public health. AQI values below 100 are generally thought of as satisfactory. When AQI values are above 100, air quality is considered to be unhealthy-at first for certain sensitive groups of people, then for everyone as AQI values get higher.

Understanding the AQI

The purpose of the AQI is to help you understand what local air quality means to your health. To make it easier to understand, the AQI is divided into six categories:

Air Quality Index (AQI) Values	Levels of Health Concern	Colors
When the AQI is in this range:	air quality conditions are:	as symbolized by this color:

0 to 50	Good	Green
51 to 100	Moderate	Yellow
101 to 150	Unhealthy for Sensitive Groups	Orange
151 to 200	Unhealthy	Red
201 to 300	Very Unhealthy	Purple
301 to 500	Hazardous	Maroon
Note: Values above 500 are considered Beyond the AQI.		

Each category corresponds to a different level of health concern. The six levels of health concern and what they mean are:

- "Good" AQI is 0 to 50. Air quality is considered satisfactory, and air pollution poses little or no risk.

- "Moderate" AQI is 51 to 100. Air quality is acceptable; however, for some pollutants there may be a moderate health concern for a very small number of people. For example, people who are unusually sensitive to ozone may experience respiratory symptoms.

- "Unhealthy for Sensitive Groups" AQI is 101 to 150. Although general public is not likely to be affected at this AQI range, people with lung disease, older adults and children are at a greater risk from exposure to ozone, whereas persons with heart and lung disease, older adults and children are at greater risk from the presence of particles in the air.

- "Unhealthy" AQI is 151 to 200. Everyone may begin to experience some adverse health effects, and members of the sensitive groups may experience more serious effects.

- "Very Unhealthy" AQI is 201 to 300. This would trigger a health alert signifying that everyone may experience more serious health effects.

- "Hazardous" AQI greater than 300. This would trigger a health warnings of emergency conditions. The entire population is more likely to be affected.

AQI colors

EPA has assigned a specific color to each AQI category to make it easier for people to understand quickly whether air pollution is reaching unhealthy levels in their communities. For example, the color orange means that conditions are "unhealthy for sensitive groups," while red means that conditions may be "unhealthy for everyone," and so on.

Air Quality Index Levels of Health Concern	Numeical Value	Meaning
Good	0 to 50	Air quality is considered satisfactory, and air pollution poses little or no risk.
Moderate	51 to 100	Air quality is acceptable; however, for some pollutants there may be a moderate health concern for a very small number of people who are unusually sensitive to air pollution.
Unhealthy for Sensitive Groups	101 to 150	Members of sensitive groups may experience health effects. The general public is not likely to be affected.
Unhealthy	151 to 200	Everyone may begin to experience health effects; members of sensitive groups may experience more serious health effects.
Very Unhealthy	201 to 300	Health alert: everyone may experience more serious health effects.
Hazardous	301 to 500	Health warnings of emergency conditions. The entire population is more likely to be affected.

Note: Values above 500 are considered Beyond the AQI.

Indices by Location

Canada

Air quality in Canada has been reported for many years with provincial Air Quality Indices (AQIs). Significantly, AQI values reflect air quality management objectives, which are based on the lowest achievable emissions rate, and not exclusively concern for human health. The Air Quality Health Index or (AQHI) is a scale designed to help understand the impact of air quality on health. It is a health protection tool used to make decisions to reduce short-term exposure to air pollution by adjusting activity levels during increased levels of air pollution. The Air Quality Health Index also provides advice on how to improve air quality by proposing behavioural change to reduce the environmental footprint. This index pays particular attention to people who are sensitive to air pollution. It provides them with advice on how to protect their health during air quality levels associated with low, moderate, high and very high health risks.

The Air Quality Health Index provides a number from 1 to 10+ to indicate the level of health risk associated with local air quality. On occasion, when the amount of air pollution is abnormally high, the number may exceed 10. The AQHI provides a local air quality current value as well as a local air quality maximums forecast for today, tonight, and tomorrow, and provides associated health advice.

1	2	3	4	5	6	7	8	9	10	+

Risk: Low (1–3) Moderate (4–6) High (7–10) Very high (above 10)

Health Risk	Air Quality Health Index	Health Messages	
		At Risk population	*General Population
Low	1–3	Enjoy your usual outdoor activities.	Ideal air quality for outdoor activities
Moderate	4–6	Consider reducing or rescheduling strenuous activities outdoors if you are experiencing symptoms.	No need to modifyyour usual outdoor activities unless you experience symptoms such as coughing and throat irritation.
High	7–10	Reduce or reschedule strenuous activities outdoors. Children and the elderly should also take it easy.	Consider reducingor rescheduling strenuous activities outdoors if you experience symptoms such as coughing and throat irritation.
Very high	Above 10	Avoid strenuous activities outdoors. Children and the elderly should also avoid outdoor physical exertion.	Reduce or reschedule strenuous activities outdoors, especially if you experience symptoms such as coughing and throat irritation.

Hong Kong

On the 30 December 2013 Hong Kong replaced the Air Pollution Index with a new index called the *Air Quality Health Index.*This index, reported by the Environmental Protection Department, is measured on a scale of 1 to 10+ and considers four air pollutants: ozone; nitrogen dioxide; sulphur dioxide and particulate matter (including PM10 and PM2.5). For any given hour the AQHI is calculated from the sum of the percentage excess risk of daily hospital admissions attributable to the 3-hour moving average concentrations of these four pollutants. The AQHIs are grouped into five AQHI health risk categories with health advice provided:

Health risk category	AQHI
Low	1
	2
	3
Medium	4
	5
	6
High	7
Very High	8
	9
	10
Serious	10+

Each of the health risk categories has advice with it. At the *low*and *moderate* levels the public are advised that they can continue normal activities. For the *high* category, children, the elderly and people with heart or respiratory illnesses are advising to reduce outdoor physical exertion. Above this (*very high* or *serious*) the general public are also advised to reduce or avoid outdoor physical exertion.

Mainland China

China's Ministry of Environmental Protection (MEP) is responsible for measuring the level of air pollution in China. As of 1 January 2013, MEP monitors daily pollution level in 163 of its major cities. The AQI level is based on the level of six atmospheric pollutants, namely sulfur dioxide (SO_2), nitrogen dioxide (NO_2), suspended particulates smaller than 10 μm in aerodynamic diameter (PM_{10}), suspended particulates smaller than 2.5 μm in aerodynamic diameter ($PM_{2.5}$), carbon monoxide (CO), and ozone (O_3) measured at the monitoring stations throughout each city.

AQI Mechanics

An individual score (Individual Air Quality Index, IAQI) is assigned to each pollutant and the final AQI is *the highest* of these six scores. The final AQI value can be calculated either per hour or per 24 hours. The concentrations of pollutants can be measured quite differently. If the AQI value is calculated hourly, then SO_2, NO_2, CO concentrations are measured as average per 24h, O_3 concentration is measured as average per hour and the moving average per 8h, $PM_{2.5}$ and PM_{10} concentrations are measured as average per hour and per 24h. If the AQI value is calculated per 24h, then SO_2, NO_2, CO, $PM_{2.5}$ and PM_{10} concentrations are measured as average per 24h, while O_3 concentration is measured as the maximum 1h average and the maximum 24h moving average. The IAQI of each pollutant is calculated according to a formula published by the MEP.

The score for each pollutant is non-linear, as is the final AQI score. Thus an AQI of 300 does not mean twice the pollution of AQI at 150, nor does it mean the air is twice as harmful. The concentration of a pollutant when its IAQI is 100 does not equal twice its concentration when its IAQI is 50, nor does it mean the pollutant is twice as harmful. While an AQI of 50 from day 1 to 182 and AQI of 100 from day 183 to 365 does provide an annual average of 75, it does *not* mean the pollution is acceptable even if the benchmark of 100 is deemed safe. Because the benchmark is a 24-hour target, and the annual average must match the annual target, it is entirely possible to have safe air every day of the year but still fail the annual pollution benchmark.

AQI and Health Implications (HJ 633—2012)

AQI	Air Pollution Level	Air Pollution Category	Health Implications	Recommended Precautions
0–50	Level 1	Excellent	No health implications.	Everyone can continue their outdoor activities normally.
51–100	Level 2	Good	Some pollutants may slightly affect very few hypersensitive individuals.	Only very few hypersensitive people should reduce outdoor activities.

101–150	Level 3	Lightly Polluted	Healthy people may experience slight irritations and sensitive individuals will be slightly affected to a larger extent.	Children, seniors and individuals with respiratory or heart diseases should reduce sustained and high-intensity outdoor exercises.
151–200	Level 4	Moderately Polluted	Sensitive individuals will experience more serious conditions. The hearts and respiratory systems of healthy people may be affected.	Children, seniors and individuals with respiratory or heart diseases should avoid sustained and high-intensity outdoor exercises. General population should moderately reduce outdoor activities.
201–300	Level 5	Heavily Polluted	Healthy people will commonly show symptoms. People with respiratory or heart diseases will be significantly affected and will experience reduced endurance in activities.	Children, seniors and individuals with heart or lung diseases should stay indoors and avoid outdoor activities. General population should reduce outdoor activities.
>300	Level 6	Severely Polluted	Healthy people will experience reduced endurance in activities and may also show noticeably strong symptoms. Other illnesses may be triggered in healthy people. Elders and the sick should remain indoors and avoid exercise. Healthy individuals should avoid outdoor activities.	Children, seniors and the sick should stay indoors and avoid physical exertion. General population should avoid outdoor activities.

India

The National Air Quality Index (AQI) was launched in New Delhi on 17 September 2014 under the Swachh Bharat Abhiyan.

The Central Pollution Control Board along with State Pollution Control Boards has been operating National Air Monitoring Program (NAMP) covering 240 cities of the country having more than 342 monitoring stations. An Expert Group comprising medical professionals, air quality experts, academia, advocacy groups, and SPCBs was constituted and a technical study was awarded to IIT Kanpur. IIT Kanpur and the Expert Group recommended an AQI scheme in 2014. While the earlier measuring index was limited to three indicators, the new index measures eight parameters. The continuous monitoring systems that provide data on near real-time basis are installed in New Delhi, Mumbai, Pune and Ahmedabad.

There are six AQI categories, namely Good, Satisfactory, Moderately polluted, Poor, Very Poor, and Severe. The proposed AQI will consider eight pollutants (PM_{10}, $PM_{2.5}$, NO_2, SO_2, CO, O_3, NH_3, and Pb) for which short-term (up to 24-hourly averaging period) National Ambient Air Quality Standards are prescribed. Based on the measured ambient concentrations, corresponding standards and likely health impact, a sub-index is calculated for each of these pollutants. The worst sub-index reflects overall AQI. Likely health impacts for different AQI categories and pollutants have also been suggested, with primary inputs from the medical experts in the group. The AQI values and corresponding ambient concentrations (health breakpoints) as well as associated likely health impacts for the identified eight pollutants are as follows:

AQI Category, Pollutants and Health Breakpoints								
AQI Category (Range)	PM_{10} (24hr)	$PM_{2.5}$ (24hr)	NO_2 (24hr)	O_3 (8hr)	CO (8hr)	SO_2 (24hr)	NH_3 (24hr)	Pb (24hr)
Good (0–50)	0–50	0–30	0–40	0–50	0–1.0	0–40	0–200	0–0.5
Satisfactory (51–100)	51–100	31–60	41–80	51–100	1.1–2.0	41–80	201–400	0.5–1.0
Moderately polluted (101–200)	101–250	61–90	81–180	101–168	2.1–10	81–380	401–800	1.1–2.0
Poor (201–300)	251–350	91–120	181–280	169–208	10–17	381–800	801–1200	2.1–3.0
Very poor (301–400)	351–430	121–250	281–400	209–748	17–34	801–1600	1200–1800	3.1–3.5
Severe (401–500)	430+	250+	400+	748+	34+	1600+	1800+	3.5+

AQI	Associated Health Impacts
Good (0–50)	Minimal impact
Satisfactory (51–100)	May cause minor breathing discomfort to sensitive people.
Moderately polluted (101–200)	May cause breathing discomfort to people with lung disease such as asthma, and discomfort to people with heart disease, children and older adults.
Poor (201–300)	May cause breathing discomfort to people on prolonged exposure, and discomfort to people with heart disease.
Very poor (301–400)	May cause respiratory illness to the people on prolonged exposure. Effect may be more pronounced in people with lung and heart diseases.
Severe (401–500)	May cause respiratory impact even on healthy people, and serious health impacts on people with lung/heart disease. The health impacts may be experienced even during light physical activity.

Mexico

The air quality in Mexico City is reported in IMECAs. The IMECA is calculated using the measurements of average times of the chemicals ozone (O_3), sulphur dioxide (SO_2), nitrogen dioxide (NO_2), carbon monoxide (CO), particles smaller than 2.5 micrometers ($PM_{2.5}$), and particles smaller than 10 micrometers (PM_{10}).

Singapore

Singapore uses the Pollutant Standards Index to report on its air quality, with details of the calculation similar but not identical to that used in Malaysia and Hong Kong The

PSI chart below is grouped by index values and descriptors, according to the National Environment Agency.

PSI	Descriptor	General Health Effects
0–50	Good	None
51–100	Moderate	Few or none for the general population
101–200	Unhealthy	Mild aggravation of symptoms among susceptible persons i.e. those with underlying conditions such as chronic heart or lung ailments; transient symptoms of irritation e.g. eye irritation, sneezing or coughing in some of the healthy population.
201–300	Very Unhealthy	Moderate aggravation of symptoms and decreased tolerance in persons with heart or lung disease; more widespread symptoms of transient irritation in the healthy population.
301–400	Hazardous	Early onset of certain diseases in addition to significant aggravation of symptoms in susceptible persons; and decreased exercise tolerance in healthy persons.
Above 400	Hazardous	PSI levels above 400 may be life-threatening to ill and elderly persons. Healthy people may experience adverse symptoms that affect normal activity.

South Korea

The Ministry of Environment of South Korea uses the Comprehensive Air-quality Index (CAI) to describe the ambient air quality based on the health risks of air pollution. The index aims to help the public easily understand the air quality and protect people's health. The CAI is on a scale from 0 to 500, which is divided into six categories. The higher the CAI value, the greater the level of air pollution. Of values of the five air pollutants, the highest is the CAI value. The index also has associated health effects and a colour representation of the categories as shown below.

CAI	Description	hideHealth Implications
0–50	Good	A level that will not impact patients suffering from diseases related to air pollution.
51–100	Moderate	A level that may have a meager impact on patients in case of chronic exposure.
101–150	Unhealthy for sensitive groups	A level that may have harmful impacts on patients and members of sensitive groups.
151–250	Unhealthy	A level that may have harmful impacts on patients and members of sensitive groups (children, aged or weak people), and also cause the general public unpleasant feelings.
251–350	Very unhealthy	A level that may have a serious impact on patients and members of sensitive groups in case of acute exposure.
351-500	hazardous	A level that may have a serious impact on patients and members of sensitive groups in case of acute exposure.

The N Seoul Tower on Namsan Mountain in central Seoul, South Korea, is illuminated in blue, from sunset to 23:00 and 22:00 in winter, on days where the air quality in Seoul is 45 or less. During the spring of 2012, the Tower was lit up for 52 days, which is four days more than in 2011.

United Kingdom

The most commonly used air quality index in the UK is the *Daily Air Quality Index* recommended by the Committee on Medical Effects of Air Pollutants (COMEAP). This index has ten points, which are further grouped into 4 bands: low, moderate, high and very high. Each of the bands comes with advice for at-risk groups and the general population.

Air pollution banding	Value	Health messages for At-risk individuals	Health messages for General population
Low	1–3	Enjoy your usual outdoor activities.	Enjoy your usual outdoor activities.
Moderate	4–6	Adults and children with lung problems, and adults with heart problems, who experience symptoms, should consider reducing strenuous physical activity, particularly outdoors.	Enjoy your usual outdoor activities.
High	7–9	Adults and children with lung problems, and adults with heart problems, should reduce strenuous physical exertion, particularly outdoors, and particularly if they experience symptoms. People with asthma may find they need to use their reliever inhaler more often. Older people should also reduce physical exertion.	Anyone experiencing discomfort such as sore eyes, cough or sore throat should consider reducing activity, particularly outdoors.
Very High	10	Adults and children with lung problems, adults with heart problems, and older people, should avoid strenuous physical activity. People with asthma may find they need to use their reliever inhaler more often.	Reduce physical exertion, particularly outdoors, especially if you experience symptoms such as cough or sore throat.

The index is based on the concentrations of 5 pollutants. The index is calculated from the concentrations of the following pollutants: Ozone, Nitrogen Dioxide, Sulphur Dioxide, PM2.5 (particles with an aerodynamic diameter less than 2.5 μm) and PM10. The breakpoints between index values are defined for each pollutant separately and the overall index is defined as the maximum value of the index. Different averaging periods are used for different pollutants.

Index	Ozone, Running 8 hourly mean (μg/m3)	Nitrogen Dioxide, Hourly mean (μg/m3)	Sulphur Dioxide, 15 minute mean (μg/m3)	PM2.5Particles, 24 hour mean (μg/m3)	PM10Particles, 24 hour mean (μg/m3)
1	0–33	0–67	0–88	0–11	0–16
2	34–66	68–134	89–177	12–23	17–33
3	67–100	135–200	178–266	24–35	34–50
4	101–120	201–267	267–354	36–41	51–58
5	121–140	268–334	355–443	42–47	59–66
6	141–160	335–400	444–532	48–53	67–75
7	161–187	401–467	533–710	54–58	76–83
8	188-213	468–534	711–887	59–64	84–91
9	214–240	535–600	888–1064	65–70	92–100
10	≥ 241	≥ 601	≥ 1065	≥ 71	≥ 101

Europe

The *Common Air Quality Index* (CAQI) is an air quality index used in Europe since 2006. In November 2017, the European Environment Agency announced the *European Air Quality Index*(EAQI) and started encouraging its use on websites and for other ways of informing the public about air quality.

CAQI

As of 2012, the EU-supported project *CiteairII* argued that the CAQI had been evaluated on a "large set" of data, and described the CAQI's motivation and definition. *CiteairII* stated that having an air quality index that would be easy to present to the general public was a major motivation, leaving aside the more complex question of a health-based index, which would require, for example, effects of combined levels of different pollutatns. The main aim of the CAQI was to have an index that would encourage wide comparison across the EU, without replacing local indices. *CiteairII* stated that the "main goal of the CAQI is not to warn people for possible adverse health effects of poor air quality but to attract their attention to urban air pollution and its main source (traffic) and help them decrease their exposure."

The CAQI is a number on a scale from 1 to 100, where a low value means good air quality and a high value means bad air quality. The index is defined in both hourly and daily versions, and separately near roads (a "roadside" or "traffic" index) or away from roads (a "background" index). As of 2012, the CAQI had two mandatory components for the roadside index, NO_2 and PM_{10}, and three mandatory components for the background index, NO_2, PM_{10} and O_3. It also included optional pollutants $PM_{2.5}$, CO and SO_2. A "sub-index" is calculated for each of the mandatory (and optional if available) components. The CAQI is defined as the sub-index that represents the worst quality among those components.

Some of the key pollutant densities in µg/m³ for the hourly background index, the corresponding sub-indices, and five CAQI ranges and verbal descriptions are as follows.

Qualitative name	Index or sub-index	Pollutant (hourly) density in µg/m³			
		NO_2	PM_{10}	O_3	$PM_{2.5}$(optional)
Very low	0–25	0–50	0–25	0–60	0–15
Low	25–50	50–100	25–50	60–120	15–30
Medium	50–75	100–200	50–90	120–180	30–55
High	75–100	200–400	90–180	180–240	55–110
Very high	>100	>400	>180	>240	>110

A separate *Year Average Common Air Quality Index* (YACAQI) is also defined, in which different pollutant sub-indices are separately normalised to a value typically near unity. For example, the yearly averages of NO_2, PM_{10} and $PM_{2.5}$ are divided by 40 μg/m^3, 40 μg/m^3 and 20 μg/m^3, respectively. The overall background or traffic YACAQI for a city is the arithmetic mean of a defined subset of these sub-indices.

United States

The United States Environmental Protection Agency (EPA) has developed an Air Quality Index that is used to report air quality. This AQI is divided into six categories indicating increasing levels of health concern. An AQI value over 300 represents hazardous air quality and below 50 the air quality is good.

Air Quality Index (AQI) Values	Levels of Health Concern	Colors
0 to 50	Good	Green
51 to 100	Moderate	Yellow
101 to 150	Unhealthy for Sensitive Groups	Orange
151 to 200	Unhealthy	Red
201 to 300	Very Unhealthy	Purple
301 to 500	Hazardous	Maroon

The AQI is based on the five "criteria" pollutants regulated under the Clean Air Act: ground-level ozone, particulate matter, carbon monoxide, sulfur dioxide, and nitrogen dioxide. The EPA has established National Ambient Air Quality Standards (NAAQS) for each of these pollutants in order to protect public health. An AQI value of 100 generally corresponds to the level of the NAAQS for the pollutant. The Clean Air Act (USA) (1990) requires EPA to review its National Ambient Air Quality Standards every five years to reflect evolving health effects information. The Air Quality Index is adjusted periodically to reflect these changes.

Computing the AQI

The air quality index is a piecewise linear function of the pollutant concentration. At the boundary between AQI categories, there is a discontinuous jump of one AQI unit. To convert from concentration to AQI this equation is used:

$$I = \frac{I_{high} - I_{low}}{C_{high} - C_{low}}(C - C_{low}) + I_{low}$$

where:

I = the (Air Quality) index,

C = the pollutant concentration,

C_{low} = the concentration breakpoint that is $\leq C$,

C_{high} = the concentration breakpoint that is $\geq C$,

I_{low} = the index breakpoint corresponding to C_{low},

I_{high} = the index breakpoint corresponding to C_{high}.

EPA's table of breakpoints is:

O_3 (ppb)	O_3 (ppb)	$PM_{2.5}$ ($\mu g/m^3$)	PM ($\mu g/m^3$)	CO (ppm)	SO_2 (ppb)	NO_2 (ppb)	AQI	AQI
C_{low} - C_{high} (avg)	C_{low} - C_{high} (avg)	C_{low} - C_{high} (avg)	C_{low} - C_{high} (avg)	C_{low} - C_{high} (avg)	C_{low} - C_{high} (avg)	C_{low} - C_{high} (avg)	I_{low} - I_{high}	Category
0-54 (8-hr)	-	0.0-12.0 (24-hr)	0-54 (24-hr)	0.0-4.4 (8-hr)	0-35 (1-hr)	0-53 (1-hr)	0-50	Good
55-70 (8-hr)	-	12.1-35.4 (24-hr)	55-154 (24-hr)	4.5-9.4 (8-hr)	36-75 (1-hr)	54-100 (1-hr)	51-100	Moderate
71-85 (8-hr)	125-164 (1-hr)	35.5-55.4 (24-hr)	155-254 (24-hr)	9.5-12.4 (8-hr)	76-185 (1-hr)	101-360 (1-hr)	101-150	Unhealthy for Sensitive Groups
86-105 (8-hr)	165-204 (1-hr)	55.5-150.4 (24-hr)	255-354 (24-hr)	12.5-15.4 (8-hr)	186-304 (1-hr)	361-649 (1-hr)	151-200	Unhealthy
106-200 (8-hr)	205-404 (1-hr)	150.5-250.4 (24-hr)	355-424 (24-hr)	15.5-30.4 (8-hr)	305-604 (24-hr)	650-1249 (1-hr)	201-300	Very Unhealthy
-	405-504 (1-hr)	250.5-350.4 (24-hr)	425-504 (24-hr)	30.5-40.4 (8-hr)	605-804 (24-hr)	1250-1649 (1-hr)	301-400	Hazardous
-	505-604 (1-hr)	350.5-500.4 (24-hr)	505-604 (24-hr)	40.5-50.4 (8-hr)	805-1004 (24-hr)	1650-2049 (1-hr)	401-500	

Suppose a monitor records a 24-hour average fine particle ($PM_{2.5}$) concentration of 12.0 micrograms per cubic meter. The equation above results in an AQI of:

$$\frac{50-0}{12.0-0}(12.0-0)+0=50,$$

corresponding to air quality in the "Good" range. To convert an air pollutant concentration to an AQI, EPA has developed a calculator.

If multiple pollutants are measured at a monitoring site, then the largest or "dominant" AQI value is reported for the location. The ozone AQI between 100 and 300 is computed by selecting the larger of the AQI calculated with a 1-hour ozone value and the AQI computed with the 8-hour ozone value.

8-hour ozone averages do not define AQI values greater than 300; AQI values of 301 or greater are calculated with 1-hour ozone concentrations. 1-hour SO_2 values do not define higher AQI values greater than 200. AQI values of 201 or greater are calculated with 24-hour SO_2 concentrations.

Real time monitoring data from continuous monitors are typically available as 1-hour averages. However, computation of the AQI for some pollutants requires averaging over multiple hours of data. (For example, calculation of the ozone AQI requires computation of an 8-hour average and computation of the $PM_{2.5}$ or PM_{10} AQI requires a 24-hour average.) To accurately reflect the current air quality, the multi-hour average used for the AQI computation should be centered on the current time, but as concentrations of future hours are unknown and are difficult to estimate accurately, EPA uses surrogate concentrations to estimate these multi-hour averages. For reporting the $PM_{2.5}$, PM_{10} and ozone air quality indices, this surrogate concentration is called the NowCast. The Nowcast is a particular type of weighted average that provides more weight to the most recent air quality data when air pollution levels are changing. There is a free email subscription service for New York inhabitants - AirNYC. Subscribers get notification about AQI values changes for selected location (eg home address), based on air quality conditions.

Pollutant Standards Index

The PSI is based on six pollutants particulate matter (PM10), fine particulate matter ($PM_{2.5}$), sulphur dioxide (SO_2), carbon monoxide (CO), ozone (O_3) and nitrogen dioxide (NO_2). For each pollutant, a sub-index is calculated from a segmented linear function that transforms ambient concentrations onto a scale extending from 0 through 500.

The breakpoints used in defining each of the six pollutant sub-indices are listed as follows:

Index Category	PSI	24-hr $PM_{2.5}$ ($\mu g/m$)^3	24-hr PM_{10} ($\mu g/m$)^3	24-hr SO_2 ($\mu g/m$)^3	8-hr CO (mg/m)^3	8-hr O_3 ($\mu g/m$)^3	1-hr NO_2 ($\mu g/m$)^3 ^
Good	0 – 50	0 – 12	0 – 50	0 – 80	0 – 5.0	0 – 118	-
Moderate	51 – 100	13 – 55	51 – 150	81 – 365	5.1 – 10.0	119 – 157	-
Unhealthy	101 – 200	56 – 150	151 – 350	366 – 800	10.1 – 17.0	158 – 235	1130
Very Unhealthy	201 – 300	151 – 250	351 – 420	801 – 1600	17.1 – 34.0	236 – 785*	1131 – 2260
Hazardous	301 – 400	251 – 350	421 – 500	1601 – 2100	34.1 – 46.0	786 – 980*	2261 – 3000
Hazardous	401 – 500	351– 500	501 – 600	2101 – 2620	46.1 – 57.5	981 – 1180*	3001– 3750

Note: *When 8-hour ozone concentration exceeds 785$\mu g/m^3$, the PSI sub-index is calculated using the 1-hour concentration; ^Sub-index for nitrogen dioxide is reported only when the 1-hour concentration equals or exceeds 1130 $\mu g/m^3$.

Each sub-index i, is calculated by using a segmented linear function that relates pollutant concentration, Xi to sub-index value, Ii. A segmented linear function consists of straight-line segments joining discrete co-ordinates (i.e. breakpoints). For pollutant i and segment j, the co-ordinates of the jth breakpoints are represented by sub-index value Ii,j and the concentration Xi,j giving the ordered pair (Xi,j , Ii,j). If the observed concentration is Xi the corresponding sub-index value Ii is calculated using the following equation over the concentration range:

Equation:

$$I_i = \frac{I_{i,j+1} - I_{i,j}}{X_{i,j+1} - X_{i,j}}\left(X_i - X_{i,j}\right) + I_{i,j}$$

For $X_{i,j \leq} X_{i,j \leq}$

where

X_i = Observed concentration for the ith pollutant

$I_{i,j}$ = PSI value for the ith pollutant and the jth breakpoint as given in the table

$I_{i,j+1}$= PSI value for the ith pollutant and the $(j+1)^{th}$ breakpoint as given in the table

$X_{i,j}$ = Concentration for the ith pollutant and j^{th} breakpoint as given in the table

$X_{i,j+1}$= Concentration for the ith pollutant and $(j+1)^{th}$ breakpoint as given in the table

Finally, the overall index is calculated as the maximum of sub-indices:

PSI = maximum $(I_1, I_2, I_3, I_4, I_5, I_6)$

Example of computation

Suppose a 24-hr $PM_{2.5}$ concentration of 40 µg/m³ is observed. Based on the table, the observed concentration of Xi = 40 µg/m³ lies between 12 and 55 µg/m³. Therefore, the computation is carried out for the first segment (j = 1). For this segment, $X_{1,1}$ = 12 µg/m³ and $X_{1,2}$ = 55 µg/m³ with corresponding sub-index values of $I_{1,1}$ = 50 and $I_{1,2}$ = 100. The computation is as follows:

$$I_i = \frac{I_{i,j+1} - I_{i,j}}{X_{i,j+1} - X_{i,j}} (X_i - X_{i,j}) + I_{i,j}$$
$$= \frac{100 - 50}{50 - 12}(40 - 12) + 50$$
$$= 83$$

Therefore, the $PM_{2.5}$ sub-index is 83. If the five other pollutant sub-indices calculated in a similar manner from concentrations were $I_{2\ (PM10)} = 48$, $I_{3\ (SO2)} = 46$, $I_{4\ (CO)} = 15$, $I_{5\ (O3)} = 45$, $I_{6\ (NO2)} = -*$, then the overall index is reported as the maximum of these values as follows:

PSI = maximum (83, 48, 46, 15, 46, - *) = 83

Note: Sub-index for nitrogen dioxide is reported only when the 1-hour concentration equals or exceeds 1130 µg/m³, which corresponds to sub-index of 200.

NowCast System

Nowcast system used by the US EPA to convert the raw pollutants readings, expressed in µg/m3 or ppb, into the AQI (scale from 0 to 500).

The concept behind the *nowcast* is to compensate the *"24 hours averaging"*, which should be used when converting concentrations to AQI. The reason for this averaging is that the AQI scale specifies that each of the Levels of Health Concern (i.e. Good, Moderate, Unhealthy) is valid under a 24 hours exposure. For example, when seeing a 188 AQI (Unhealthy), one need to read it as *"if I stay out for 24 hours, and the AQI is 188 during those 24 hours, then the health effect is Unhealthy"*. This is quite different from saying that *"if the AQI reported now is 188, then the health effect is Unhealthy"*.

The problem is that the 24 hours averaging is a very bad idea, and should abolished for at least those two reasons:

- First, the dynamic of Air Pollution is such that wind than completely clean the air in less than 30 minutes. This phenomenon is frequently seen in Beijing with the strong north winds able to bring the $PM_{2.5}$ AQI from more than 300 to less than 50 in less than one hour. When this happens, no one wants to wait for 24 hours before knowing that the Air Quality is good (and go out for a walk to enjoy the fresh air).

- The second reason is when the Air Quality suddenly gets worse. One famous case is the Indonesian wildfire causing the Singapore Smog when the winds are heading to the north, under which circumstances the AQI can go from below 50 to more than 150 is just one hour. As a matter of fact, we have had many requests from asthmatic/sensitive people when Singapore was still only providing 24 hours average readings.

This is for those reasons that the US EPA introduced the *nowcast* system: It is an alternative conversion formula used to counter balance the need for averaging under changing Air Quality conditions.

NowCast is an air quality data smoothing algorithm that puts an emphasis on recent values when measurements are unstable, and approaches a long-term (*e.g.* 12-hour) average when measurements are stable.

The NowCast is designed to be responsive to rapidly changing air quality conditions, such as during a wildfire. The NowCast calculation uses longer averages during periods of stable air quality and shorter averages when air quality is changing rapidly. The NowCast allows AirNow's current conditions to align more closely with what people are actually seeing or experiencing. This gives people information they can use to protect their health when air quality is poor – and help them get outdoors and get exercise when air quality is good.

Calculating the NowCast for PM

Use the past 12 hours of PM measurements in micrograms per cubic meter (µg/m³):

Select the minimum and maximum PM measurements

1. Subtract the minimum measurement from the maximum measurement to get the range.

2. Divide the range by the maximum measurement in the 12 hour period to get the scaled rate of change.

4. Subtract the scaled rate of change from 1 to get the weight factor. The weight factor must be between .5 and 1. The minimum limit approximates a 3-hour average.

5. If the weight factor is less than 0.5, then set it equal to 0.5.

6. Multiply each hourly measurement by the weight factor raised to the power of the number of hours ago the value was measured (for the current hour, the factor is raised to the zero power).

7. Compute the NowCast by summing the products from Step 6 and dividing by the sum of the weight factor raised to the power of the number of hours ago each value was measured.

8. Convert this value to an AQI.

Let $c_1, c_2, ...c_{12}$ represent the hourly PM concentrations for the most recent 12-hour period, with c_1 the most recent hourly value, and let c_{min} and c_{max} represent the minimum and maximum hourly concentration for the 12-hour period.

Define:

$$w^* = \frac{c_{min}}{c_{max}}$$

and let

$$w = \begin{cases} w^* & \text{if } w^* > \dfrac{1}{2}, \\ \dfrac{1}{2} & \text{if } w^* \leq \dfrac{1}{2}. \end{cases}$$

With these definitions the PM NowCast is given by:

$$NowCast = \frac{\sum_{i=1}^{12} w^{i-1} c_i}{\sum_{i=1}^{12} w^{i-1}}.$$

For the special case where there is no variability in the hourly values, $c_{min} = c_{max}$, $w = 1$, and the NowCast reduces to the twelve-hour average:

$$NowCast = \frac{\sum_{i=1}^{12} c_i}{12}.$$

For the special case where w=1/2:

$$NowCast = \frac{c_1 + \left(\dfrac{1}{2}\right) c_2 + \ldots + \left(\dfrac{1}{2}\right)^{11} c_{12}}{1 + \dfrac{1}{2} + \left(\dfrac{1}{2}\right)^2 + \ldots + \left(\dfrac{1}{2}\right)^{11}}$$

But $1/(1-x)=1 + x + x^2 + \ldots$, $x < 1$, so to a good approximation, when $w = \frac{1}{2}$:

$$NowCast = \frac{1}{2} c_1 + \left(\frac{1}{2}\right)^2 c_2 + \ldots + \left(\frac{1}{2}\right)^{12} c_{12}$$

Because the most recent hours of data are weighted so heavily in the NowCast when PM levels are changing, EPA does not report the NowCast when data is missing for c_1 or c_2.

Calculating the NowCast for Ozone

The ozone NowCast replaces the ozone surrogate (slope/intercept method) for real time reporting.

The ozone NowCast calculation is similar to the PM NowCast calculation. It does have some slight differences below:

1. Compute the concentrations range (max-min) over the last 8 hours.

2. Divide the range by the maximum concentration in the 8 hour periodto obtain the scaled rate of change.

3. Compute the weight factor by subtracting the scaled rate from 1. There is no minimum weight factor (PM uses a minimum weight factor of 0.5). There is a maximum weight factor of 1.

4. Multiply each hourly concentrations by the weight factor raised to the power of how many hours ago the concentration was measured (for the current hour, the factor is raised to the zero power).

5. Compute the NowCast by summing these products and dividing by the sum of the weight factors raised to the power of how many hours ago the concentration was measured.

Let $c_1, c_2, \ldots c_8$ represent the hourly ozone concentrations for the most recent 8-hour period, with c_1 the most recent hourly value, and let c_{min} and c_{max} represent the minimum and maximum hourly concentration for the 8-hour period.

Define:

$$w = \frac{c_{min}}{c_{max}}$$

With these definitions the Ozone NowCast is given by

$$NowCast = \frac{\sum_{i=1}^{8} w^{i-1} c_i}{\sum_{i=1}^{8} w^{i-1}}.$$

Note: USEPA generally truncates ozone values to whole PPB, so a value of 13.78 PPB would use a value of 13.0 in the NowCast calculation, and a value of 0.01378 PPM would use a value 0.013 PPM.

Example

Consider a day when the hourly average $PM_{2.5}$ concentration is zero for all hours of the day, except for a single hour from noon to 1 pm, where a monitor records a concentration pulse of 71 micrograms per cubic meter ($\mu g/m^3$). According to the equation above, the Nowcast is 71/2 $\mu g/m^3$=35.5 $\mu g/m^3$ the hour after the pulse, two hours later it is 71/4 $\mu g/m^3$=17.8 $\mu g/m^3$ and three hours later it is 71/8 $\mu g/m^3$= 8.9 $\mu g/m^3$. To calculate the corresponding AQI values, each NowCast concentration is substituted into the AQI equation in place of the 24-hour average $PM_{2.5}$ concentration:

$$I = \frac{C - C_{low}}{C_{high} - C_{low}} (I_{high} - I_{low}) + I_{low}$$

where:

I = the AQI,

C = the 24-hour average PM$_{2.5}$ pollutant concentration,

C_{low} = the concentration breakpoint that is $\leq C$,

C_{high} = the concentration breakpoint that is $\geq C$,

I_{low} = the index breakpoint corresponding to C_{low},

I_{high} = the index breakpoint corresponding to C_{high}.

and:

C_{low}	C_{high}	I_{low}	I_{high}	Category
0	12.0	0	50	Good
12.1	35.4	51	100	Moderate
35.5	55.4	101	150	Unhealthy for Sensitive Groups
55.5	150.4	151	200	Unhealthy
150.5	250.4	201	300	Very Unhealthy
250.5	350.4	301	400	Hazardous
350.5	500.4	401	500	Hazardous

Thus, the three NowCast concentrations correspond to air quality indices of 101, (AQI Color Code Orange, Air Quality: Unhealthy for Sensitive Groups), 63 (AQI Color Code Yellow, Air Quality: Moderate), and 37 (AQI Color Code Green, Air Quality: Good) respectively. After the day is over and all of the hourly data is available, the AQI for the day is calculated from the 24-hr average; 71/24 µg/m³= 3.0 µg/m³, an AQI of 12 (Color Code Green, Air Quality: Good). EPA has developed a calculator to compute the PM NowCast, AQI, and AQI category and color from the most recent 12 hours of monitoring data.

Beta Attenuation Monitoring

Instruments such as Beta Attenuation Monitors (BAMs) provide real-time PM$_{2.5}$ mass concentration data and are now widely used in compliance monitoring networks. These devices collect the PM on a filter tape in 1-hour samples. In this work, the use of BAM filter spots for chemical speciation was investigated. Filter tapes from several sites in California were analyzed for a series of chemical species including elements, ions, organic and elemental carbon, and molecular markers. A major issue was the blank values in the filter tape. Based on blank spot analyses, it was determined that measurement of organic and elemental carbon (OC/EC) and elements were infeasible. A total of 22 BAM samples (each comprising 24 1-hour spots) from 12 sites were analyzed for ions and black carbon (BC). Additionally, 336 1-hour spots were composited to analyze for molecular markers (MM). Measurements of ions and BC at each site appear to have been underestimated likely due to volatilization losses. MM measurements in these 336 BAM filter spots suggest that organic speciation of BAM filters could be a viable method for measuring useful marker species. Statistical analysis was conducted by grouping samples into classes with mass concentrations greater and less than 80 μg m^{-3}. Measurement of Delta-C concentrations (i.e., Delta C = $B_{C370\ nm} - BC_{880\ nm}$) in these two groups of samples revealed that most of the high PM$_{2.5}$ concentration days PM$_{2.5}$ greater than 80 μg m^{-3}) were likely not substantially impacted by biomass combustion particles. A major portion of the samples analyzed in this work with high concentrations of PM$_{2.5}$ were inorganic species such as ammonium sulfate, ammonium nitrate and other ions. Results suggested that some useful information may be derived from BAM tape analyses. However, different filter substrates and sample handling and storage practice might make a wider range of analyses possible.

Different approaches measure different properties of atmospheric particulates, and therefore care must be taken before selecting a monitoring method or attempting to compare the results of different methods.

PM$_{10}$

PM$_{10}$ is atmospheric particulate matter less than or equal to 10 micrometres (μm) in diameter. This is the fraction of atmospheric particulates that are small enough to penetrate deep into the human lung. To monitor PM$_{10}$, the sample air enters a size-selective inlet which has at least 50 per cent efficiency cut-off at a 10 μm aerodynamic diameter. The resulting air stream contains particulate matter generally less than 10 μm (with a small proportion of particulate matter greater than 10 μm).

PM$_{10}$ can arise from a wide range of sources, but can generally be separated into three categories:

- primary combustion particulates – produced directly from combustion, such as domestic heating, road transport, power stations and industrial processes;

- secondary particulates – aggregates in the atmosphere following their release as gases (include nitrates and sulphates);

- coarse particulates – from non-combustion sources such as re-suspended road dust, construction work, mineral extraction, wind-blown dust and soil, and sea salt.

PM$_{2.5}$

PM$_{2.5}$ is particulate matter less than or equal to 2.5 μm in diameter. PM$_{2.5}$ is emitted from primary combustion processes and requires the appropriate size-selective inlet for sampling.

Current research indicates that it has a greater effect on health than PM_{10}. The AAQG include a monitoring guideline of 25 µg/m³ as a 24-hour average for $PM_{2.5}$. This is equivalent to the World Health Organization (WHO) ambient air quality guideline for $PM_{2.5}$ as a 24-hour average. WHO further provides an ambient air quality guideline of 10 µg/m³ as an annual average for $PM_{2.5}$.

Standards for monitoring $PM_{2.5}$ include:

- AS/NZS 3580.9.10:2006, *Methods for sampling and analysis of ambient air – Determination of particulate matter – $PM_{2.5}$ low volume sampler – Gravimetric method;*

- US Code of Federal Regulations Title 40, Part 50 Appendix L, *Reference method for the determination of fine particulate matter as $PM_{2.5}$ in the atmosphere.*

Air Pollution Sensor

The air quality sensor is part of the air conditioning system. It measures pollutants, in the form of oxidisable or reducible gases, in the air outside your car. Oxidisable gases include carbon monoxide, hydrocarbons (vapours from benzene or petrol) and other incompletely burnt fuel components. If the quality of the outside air drops, the control system activates the Climatronic air recirculation mode, stopping polluted air from coming in and maintaining the quality of air inside the car.

The emergence of small-scale air quality sensors has led to a significant paradigm shift in the approach to measuring air quality beyond those afforded by traditional methods that use large, stationary, and expensive analyzers. These sensors are, not only small, but also can be portable, providing data in near-real time at relatively low costs and using low amounts of power. As a result, sensors allow air quality to be measured with unprecedented temporal and spatial resolution, transforming the way we understand our environment. Sensor-based measurement devices are being used by scientists looking to better characterize air quality and its environmental and health impacts, as well as the emergence of Citizen Science, empowering individuals to communities to reduce their risk from air pollution. Key factors enabling this widespread use include progress in miniaturized electronics and microfabrication, allowing for easy and inexpensive mass production. Sensors are currently available or being developed to measure ambient concentrations of air pollutants found in air, e.g., NO, NO_2, O_3, CO_2, CO, CH_4, VOC, organic species, as well as particulate matter (PM) mass in one or more size ranges and components of PM, e.g., black carbon. However, many of the commercially available sensors have not been thoroughly evaluated and, currently, a significant fraction perform poorly relative to reference methods.

Air Quality Forecasting

Air Quality Forecasting - also refered as 'Atmospheric Dispersion Modeling' is the *art* of simulating how air pollutants (eg $PM_{2.5}$ or Ozone) disperse is the ambient atmosphere. The result of the simulation gives the ambient concentration for each air pollutants, from which the Air Quality Index can be calculated.

The Need for Air Quality Forecasts

A system for forecasting future air quality cannot, by itself, solve the problems described above. Forecasts, if they are reliable and sufficiently accurate, can however play an important role as part of an air quality management system working in concert with more traditional emissions-based approaches. The applications of air quality forecasts fall into the following broad areas:

- *Health Alerts* – Many cities currently provide warnings to the public when air pollution levels exceed specified levels. The more reliable the forecast is the more effective it is. These warnings are directed at specific populations that are particularly sensitive to air pollution (e.g., asthmatics). Interest in finding innovative ways to protect these individuals has heightened in recognition of lack of a discernable health threshold for exposure to ozone or fine particles, which implies that no level of emissions reduction will protect all individuals.

- *Supplementing Existing Emission Control Programs* – In many parts of the country, the air quality standards are exceeded only infrequently, a few days out of the year. The availability of reliable air pollution forecasts affords local environmental regulators the option of "on demand" or intermittent emission reductions on those days, thus avoiding the high cost of continuous emission control. This approach is currently being successfully employed in several areas of the country and could be expanded were reliable forecasts available. Many cities also offer free access to public transportation on "ozone alert" days to reduce automobile emissions. The accuracy of these forecasts is critical due to the high cost associated with these programs.

- *Operational Planning* – Regional haze can impair and even endanger activities such as private and commercial aviation. Aerial photography and visits to many National Parks are significantly impacted by the presence of haze. A reliable visibility forecast could improve safety and efficiency by permitting the scheduling of these activities during the most favorable periods. The U.S. Forest Service (U.S.F.S.) is planning a 10-fold increase in prescribed burns. Since these activities are regulated under the Clean Air Act, the U.S.F.S. will have to demonstrate to local regulators that they can schedule these burns so that no National Ambient Air Quality Standards will be violated, requiring some form of air quality forecast.

- *Emergency Response* – Wildfires consumed more than 4 million acres of forest in the United States during 2000. The vast amount of smoke generated by these burning forests affects the visibility in the area that can cause accidents, increase traffic congestion and even jeopardize aviation safety. The availability of reliable smoke forecasts offers rerouting options for automobiles and air traffic to reduce the possibility of accidents. The information provided by these forecasts can also provide an initial assessment of the impact of these wildfires.

Air Quality Forecasting Techniques

A wide variety of techniques, ranging from the simple to the complex, have been used to produce air quality forecasts. To date, most of these efforts have focused on producing 1- to 3-day ozone forecasts. The techniques that have been used to produce these forecasts are described in a recent report [U.S. EPA, 1999]. The techniques used to forecast ozone concentrations are representative of those that can, or could, be used for other pollutants. They fall into three broad categories:

Climatology – The use of climatology to predict air quality is based on the assumption that the past is a good predictor of the future. This approach relies on the association of elevated pollution levels with specific meteorological conditions. The application can be as simple as assuming persistence (i.e., if pollution levels are high today they will also be high tomorrow) or can involve the development of complex weather typing schemes (i.e., identifying recurring weather patterns that are accompanied by high pollution levels) to forecast air quality. These approaches are usually used to predict exceedances of specific thresholds not ambient concentrations. These approaches do, however, have the advantage of being reasonably simple and inexpensive to implement and operate.

Statistical Methods – The association between specific meteorological parameters and air quality can be quantified using a variety of statistical techniques. In fact, these are probably the most common techniques in use for ozone forecasting. In their survey,

EPA, 1999 has identified three statistical approaches that are in use:

- Classification and Regression Tree (CART) – This technique uses specialized software to identify those variables (meteorological or air quality) that are most strongly correlated with ambient pollution levels. These variables are then used to predict future pollution levels based on current air quality and forecasted meteorology.

- Regression Analysis – The association between pollutant levels and meteorological and aerometric variables can be quantified by analyzing historical data sets using standard statistical analysis packages. The resultant multi-variant linear regression equation can be used to forecast future pollution levels.

- Artificial Neural Networks – Another way of analyzing historical data is to identify atmospheric parameters that influence air quality and quantify that association through the application of adaptive learning and pattern recognition techniques, such as neural networks. Neural networks are intended to mimic the way the human brain recognizes recurring patterns. Networks have been developed that identify weather patterns that are associated with elevated ozone levels. Presumably, the same technique could be applied to other pollutants.

These approaches, while more complex than the ones discussed in the previous group, are reasonably simple to develop and use, requiring only modest computing resources and specialized knowledge.

Three Dimensional (3-D) Models – Although the techniques described above have many strong points, they have a common weakness. They assume a certain amount of stability in terms of the processes that affect air quality. Any change in emissions or climate (short and long-term) will serve to diminish the skill of these techniques. One way around this problem is to employ a more deterministic approach to the prediction of air quality. Deterministic 3-D air quality models seek to mathematically represent all of the important processes that affect ambient pollution levels. These models are actually comprised of several submodels that work together to simulate the emission, transport, and transformation of air pollution. Examples of submodels include:

- Emissions models – These models simulate the time-dependent, spatially- distributed emissions of the pollutant in question, and/or (in the case of secondary pollutants such as O_3) its precursors, from both anthropogenic and natural sources.

- Meteorological models – These models forecast meteorological conditions that determine transport and mixing and influence chemistry (solar intensity, temperature, humidity, etc.), emissions (e.g. temperature), and deposition.

Trajectory models use the 3-D meteorology from these models in consort with emissions data to forecast ambient levels of reasonably unreactive pollutants like dust and smoke.

- Chemical models – These models use fundamental chemical kinetic rate parameters, spectroscopic properties, and thermodynamic relationships to simulate the transformation of primary (emitted) pollution into secondary pollution, including the composition and morphology (size distribution and optical properties) of aerosols.

Three-dimensional air quality models are classified as being either Lagrangian or Eulerian depending on the method used to simulate the time-dependent distribution of pollution concentrations. Lagrangian models follow individual air parcels over time using the meteorological field to advect and disperse the pollutants. This approach results in a computationally efficient system. However, it is difficult to properly characterize the interaction of a large number of individual sources when nonlinear chemistry is involved. Eulerian models use fixed grids (vertically and horizontally) and solve the appropriate chemical equations simultaneously in all cells, including exchange of pollutants between cells. Typically the computational requirements are reduced through the use of nested grids, with a coarse grid used over rural areas (where concentrations tend to be reasonably homogenous) and a finer grid used over urban areas (where concentration gradients tend to be more pronounced). These models can also accommodate a plume-in-grid treatment by performing a semi-Lagrangian calculation for large point sources (e.g., power plants) during the early stages of plume dilution. These models can produce three-dimensional concentration fields for several pollutants but require significant computational power and expertise.

Virtually all of the techniques described above start with a meteorological forecast. Therefore, the reliability of the air quality forecast is dependent on the reliability of the weather forecast. Weather forecasters use a number of tools to predict tomorrow's weather. Local forecasters will typically use the output from several different models in combination with local knowledge and experience to produce an accurate forecast. The same must be true for an air pollution forecast. A skilled forecaster will combine several of the techniques described to ensure that the prediction is as accurate as possible.

Elements of an Air Quality Forecasting System

As with a weather forecasting system, an air quality forecasting system must contain a compatible combination of two components. These components are a suite of predictive models/techniques tailored to specific needs of the customer community and an observation network capable of providing real-time measurements of atmospheric composition needed to initialize the models and evaluate the quality of the forecast.

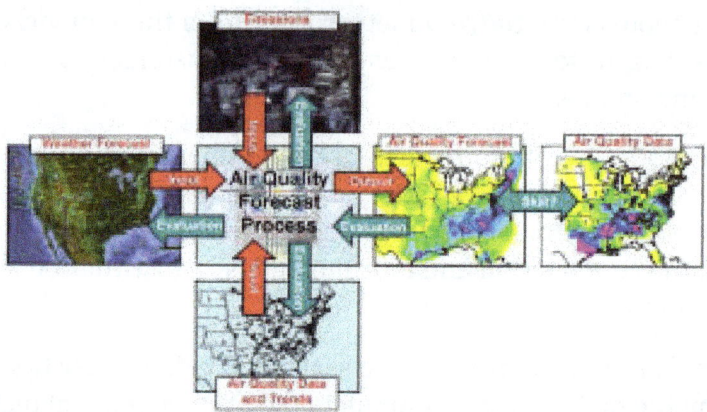

Schematic showing the interrelation among the main elements of an air quality forecasting system. A well designed forecast system includes both a process for producing the forecast and an observing system to evaluate the quality of the forecast and identify areas where improvement is needed.

A comprehensive observing system must be an integral part of any forecast system that is developed to predict future air quality. It is only by comparing actual and forecast pollution levels that we can assign confidence to future forecasts and identify areas where improvements are needed.

Traditionally the evaluation of air quality models (whatever their form) has fallen into two broad categories:

- *Operational Evaluation* – An operational evaluation involves a direct comparison between the forecast pollutant fields and the observed pollutant distribution. For example, in the case of ozone the model predictions would be compared against the concentrations measured in the regulatory network, and some skill score calculated based on a point-by-point comparison at the monitoring sites. The existing regulatory network for ozone is well suited to such an evaluation. Unfortunately the situation for other pollutants (e.g., fine particles) is not as good since those monitoring networks are less dense and the instruments not as sensitive, or in some cases as selective, as is desired.

- *Diagnostic Evaluation* – An operational evaluation will tell you how close the model came to the correct answer; a diagnostic evaluation will tell you if you got the right answer for the right reason. As the name implies, a diagnostic evaluation requires the measurement of parameters (both meteorological and chemical) that control pollutant formation and distribution, not just the concentration of the pollutant that is being forecasted. To perform a diagnostic evaluation, concentrations of pollutant precursors and key intermediates need to be tracked to evaluate the performance of the emissions model and the chemical processor (if one is used), while meteorological parameters such as mixing height and winds aloft will aid in the evaluation of the meteorological processor.

Air Quality Forecasting as a Way to Improve Understanding

As techniques for forecasting air quality improve and their use expands, we should not overlook the opportunity to use this process to improve our understanding of the processes that control the formation and distribution of air pollution. The meteorological research community has benefited enormously from the experience gained through an operational forecasting enterprise. The ongoing evaluation of the daily weather forecasts is used to identify areas of insufficient understanding and guide research.

The same opportunity exists for advancing the understanding of atmospheric processes that control ambient pollution levels. By evaluating the success of forecasts produced by different techniques we have the opportunity not only to evaluate the relative merits of these techniques but also to test our knowledge of key processes and identify areas where more information is needed.

Chapter 5

Air Pollution Mitigation

Various technologies and strategies are available for the reduction of air pollution. The aim of this chapter is to explore the various air pollution mitigation technologies, such as the use of thermal oxidizer, emissions trading, dust collection system, diesel particulate filter, etc.

Thermal Oxidizer

Thermal Oxidisers are a class of pollution control devices that use the combustion process to destroy hydrocarbons, volatile organic compounds (VOC), Hydrogen Sulphide, and other compounds as well as smoke particles and odour, hence they are sometimes called afterburners, fume incinerators or tail gas incinerators.

The design of an incineration system is dependent on the pollutant concentration in the waste gas stream, type of pollutant, presence of other gases, level of oxygen, stability of processes vented to the system, and degree of control required. Important design factors include temperature (a temperature high enough to ignite the organic constituents in the waste gas stream), residence time (sufficient time for the combustion reaction to occur), and turbulence or mixing of combustion air with the waste gas. Time, temperature, degree of mixing, and sufficient oxygen concentration govern the completeness of the combustion reaction. Of these, only temperature and oxygen concentration can be significantly controlled after construction. Residence time and mixing are fixed by oxidizer design, and flow rate can be controlled only over a limited range.

The rate at which VOC compounds, volatile HAP, and CO are oxidized is greatly affected by temperature; the higher the temperature, the faster the oxidation reaction proceeds. Because inlet gas concentrations are well below the lower explosive limit (LEL) to prevent preignition explosions in ducting the stream from the process to the oxidizer, the gas must be heated with auxiliary fuel above the autoignition temperature. Thermal destruction of most organics occurs at combustion temperatures between 800°F and 2000°F. Residence time is equal to the oxidizer chamber volume divided by the total actual flow rate of flue gases (waste gas flow, added air, and products of combustion). A residence time of 0.2 to 2.0 seconds, a length-todiameter ratio of 2 to 3 for the chamber dimensions, and an average gas velocity of 10 to 50 feet per second are common. Thorough mixing is necessary to ensure that all waste and fuel come in contact with oxygen.

Because complete mixing generally is not achieved, excess air/oxygen is added (above stoichiometric or theoretical amount) to ensure complete combustion.

Normal operation of a thermal oxidizer should include a fixed outlet temperature or an outlet temperature above a minimum level. A variety of operating parameters that may be used to indicate good operation include: inlet and outlet VOC concentration, outlet combustion temperature, auxiliary fuel input, fuel pressure (magnehelic gauge), fan current (ammeter), outlet CO concentration, and outlet O_2 concentration.

Technologies

Direct Fired Thermal Oxidizer - Afterburner

The simplest technology of thermal oxidation is direct-fired thermal oxidizer. A process stream with hazardous gases is introduced into a firing box through or near the burner and enough residence time is provided to get the desired destruction removal efficiency (DRE) of the VOCs. Most direct-fired thermal oxidizers operate at temperature levels between 980 °C (1,800 °F) and 1,200 °C (2,190 °F) with air flow rates of 0.24 to 24 standard cubic meters per second.

Direct-fired thermal oxidizer using landfill gas as fuel

Also called afterburners in the cases where the input gases come from a process where combustion is incomplete, these systems are the least capital intensive, and can be integrated with downstream boilers and heat exchangers to optimize fuel efficiency. Thermal Oxidziers are best applied where there is a very high concentration of VOCs to act as the fuel source (instead of natural gas or oil) for complete combustion at the targeted operating temperature.

Regenerative Thermal Oxidizer (RTO)

One of today's most widely accepted air pollution control technologies across industry is a regenerative thermal oxidizer, commonly referred to as a RTO. RTOs use a ceramic bed which is heated from a previous oxidation cycle to preheat the input gases to partially oxidize them. The preheated gases enter a combustion chamber that is heated by an external fuel source to reach the target oxidation temperature which is in the range between 760 °C (1,400 °F) and 820 °C (1,510 °F). The final temperature may be as high as 1,100 °C (2,010 °F) for applications that require maximum destruction. The air flow rates are 2.4 to 240 standard cubic meters per second.

Regenerative thermal oxidizer (RTO) that is 17000 standard cubic feet per minute, or SCFM for short.

Control center with a programmable logic controller for a RTO.

RTOs are very versatile and extremely efficient – thermal efficiency can reach 95%. They are regularly used for abating solvent fumes, odours, etc. from a wide range of industries. Regenerative Thermal Oxidizers are ideal in a range of low to high VOC concentrations up to 10 g/m³ solvent. There are currently many types Regenerative Thermal Oxidizer on the market with the capabitlity of 99.5+% Volatile Organic Compound (VOC) oxidisation or destruction efficiency. The ceramic heat exchanger(s) in the towers can be designed for thermal efficiencies as high as 97+%.

Ventilation Air Methane Thermal Oxidizer (VAMTOX)

Ventilation air methane thermal oxidizers are used to destroy methane in the exhaust air of underground coal mine shafts. Methane is a greenhouse gas and, when oxidized via thermal combustion, is chemically altered to form CO_2 and H_2O. CO_2 is 25 times less potent than methane when emitted into the atmosphere with regards to global warming. Concentrations of methane in mine ventilation exhaust air of coal and trona mines are very dilute; typically below 1% and often below 0.5%. VAMTOX units have a system of valves and dampers that direct the air flow across one or more ceramic filled bed(s). On start-up, the system preheats by raising the temperature of the heat exchanging ceramic material in the bed(s) at or above the auto-oxidation temperature of methane 1,000 °C (1,830 °F), at which time the preheating system is turned off and mine exhaust air is introduced. Then the methane-filled air reaches the preheated bed(s), releasing the heat from combustion. This heat is then transferred back to the bed(s), thereby maintaining the temperature at or above what is necessary to support auto-thermal operation.

Thermal Recuperative Oxidizer

A less commonly used thermal oxidizer technology is a thermal recuperative oxidizer. Thermal recuperative oxidizers have a primary and/or secondary heat exchanger within the system. A primary heat exchanger preheats the incoming dirty air by recuperating heat from the exiting clean air. This is done by a shell and tube heat exchanger or a plate heat exchanger. As the incoming air passes on one side of the metal tube or plate, hot clean air from the combustion chamber passes on the other side of the tube or plate and heat is transferred to the incoming air through the process of conduction using the metal as the medium of heat transfer. In a secondary heat exchanger the same concept applies for heat transfer, but the air being heated by the outgoing clean process stream is being returned to another part of the plant – perhaps back to the process.

Biomass Fired Thermal Oxidizer

Biomass, such as wood chips, can be used as the fuel for a thermal oxidizer. The biomass is then gasified and the stream with hazardous gases is mixed with the biomassgas in a firing box. Sufficient turbulence, retention time, oxygen content and temperature will ensure destruction of the VOC's. Such biomass fired thermal oxidizer has been installed at Warwick Mills, New Hampshire. The inlet concentrations are between 3000-10.000 ppm VOC. The outlet concentration of VOC are below 3 ppm, thus having a VOC destruction efficiency of 99.8%-99.9%.

Flameless Thermal Oxidizer (FTO)

In a flameless thermal oxidizer system waste gas, ambient air, and auxiliary fuel are premixed prior to passing the combined gaseous mixture through a preheated inert ceramic media bed. Through the transfer of heat from the ceramic media to the gaseous mixture the organic compounds in the gas are oxidized to innocuous byproducts, i.e., carbon dioxide (CO_2) and water vapor (H_2O) while also releasing heat into the ceramic media bed.

The gas mixture temperature is kept below the lower flammability limit based on the percentages of each organic species present. Flameless thermal oxidizers are designed to operate safely and reliably below the composite LFL while maintaining a constant operating temperature. Waste gas streams experience multiple seconds of residence time at high temperatures leading to measured destruction removal efficiencies that exceed 99.9999%. Premixing all of the gases prior to treatment eliminates localized high temperatures which leads to thermal NOx typically below 2 ppmV. Flameless thermal oxidizer technology was originally developed at the U.S. Department of Energy to more efficiently convert energy in burners, process heaters, and other thermal systems.

Catalytic Oxidizer

Catalytic oxidizer (also known as catalytic incinerator) is another category of oxidation systems that is similar to typical thermal oxidizers, but the catalytic oxidizers use a catalyst to promote the oxidation. Catalytic oxidation occurs through a chemical reaction between the VOC hydrocarbon molecules and a precious-metal catalyst bed that is internal to the oxidizer system. A catalyst is a substance that is used to accelerate the rate of a chemical reaction, allowing the reaction to occur in a normal temperature range between 340 °C (644 °F) and 540 °C (1,004 °F).

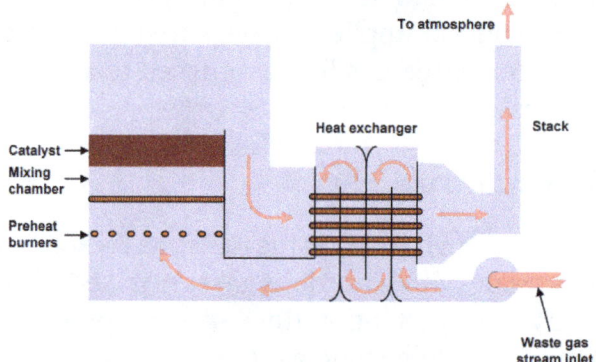

Schematic of Recuperative Catalytic Oxidizer

Regenerative Catalytic Oxidizer (RCO)

The catalyst can be used in a Regenerative Thermal Oxidizer (RTO) to allow lower operating temperatures. This is also called Regenerative Catalytic Oxidizer or RCO. For example, the thermal ignition temperature of carbon monoxide is normally 609 °C (1,128 °F). By utilizing a suitable oxidation catalyst, the ignition temperature can be reduced to around 200 °C (392 °F). This can result in lower operating costs than a RTO. Most systems operate within the 260 °C (500 °F) to 1,000 °C (1,830 °F) degree range. Some systems are designed to operate both as RCOs and RTOs. When these systems are used special design considerations are utilized to reduce the probability of overheating (dilution of inlet gas or recycling), as these high temperatures would deactivate the catalyst, e.g. by sintering of the active material.

Recuperative Catalytic Oxidizer

Catalytic oxidizers can also be in the form of recuperative heat recovery to reduce the fuel requirement. In this form of heat recovery, the hot exhaust gases from the oxidizer pass through an heat exchanger to heat the new incoming air to the oxidizer.

Ventilation Air Methane Thermal Oxidizer

Ventilation air methane thermal oxidizers (or VAMTOX) are a type of processing equipment used for greenhouse gas abatement related to underground mining operations that destroys gaseous methane at a high temperature.

Principle

Ventilation Air Methane Thermal Oxidizers are used to destroy methane in the exhaust air of underground coal mine shafts. Methane is a greenhouse gas that burns to form carbon dioxide (CO_2) and water vapour. Carbon dioxide is 25 times less potent than methane when emitted into the atmosphere with regards to global warming. Concentrations of methane in ventilation exhaust air of coal and trona mines are very dilute; typically below 1% and often below 0.5%. Flow rates are so high that ventilation air methane constitutes the largest source of methane emissions at most mines. This methane emission wastes energy and contributes significantly to global greenhouse gas (GHG) emissions.

Ventilation Air Methane (VAM) is a Greenhouse Gas (GHG) released by coal or mineral mining operations, it is also commonly known as Coal Mine Methane (CMM). The air pollutants from this process have a global warming potential 21 times greater than carbon dioxide, which is another well known GHG. As alternative energy technologies have developed, methane is increasingly used to generate electricity and heat. Landfills and mines represent two of the most abundant sources, however reports indicate that more than 50% of all VAM is exhausted from mine ventilation systems and remains unutilized.

Mining operations can profit by not venting the methane to atmosphere if they incorporate the proper VAM oxidation technology for air purification and combine it with a heat-to-power system for electricity production. These projects can often be funded without using extensive capital resources when the carbon credit and electrical power potential is taken into account.

Based on the general process conditions, a Regenerative Thermal Oxidizer (RTO) is clearly the best abatement device for VAM but it is certainly not a standard RTO application. These systems must be built conservatively and with great flexibility to ensure continuous operation. Most mine operators demand that the emission control system and related hooding for collection have zero impact on the operation and employee safety.

The sheer volume of air coming from the mine shafts is massive and can vary greatly. Therefore VAM oxidizers must handle the large volume / low concentration stream, yet be capable of adapting to emission and flow spikes.

Regenerative Thermal Oxidizer

Air Pollution Control using a Regenerative Thermal Oxidizer(RTO) is most appropriate in process applications with lower concentrations (up to 10% LEL) of Volatile Organic

Compounds (VOC's) and air flows higher than 5000 SCFM. The chief benefit of installing a Regenerative Thermal Oxidizer Media System is lower operating costs when compared to other emission control technologies. High energy efficiency, around 95%, is achieved by recovering and re-using the excess heat energy which is created in the process of combusting the organic materials contained in the air stream. It is possible to obtain destruction efficiency in excess of 99% in some cases.

Pollutant Destruction Phase

1. The process begins with a high-pressure supply fan that forces exhaust fumes into the Regenerative Thermal Oxidizer.

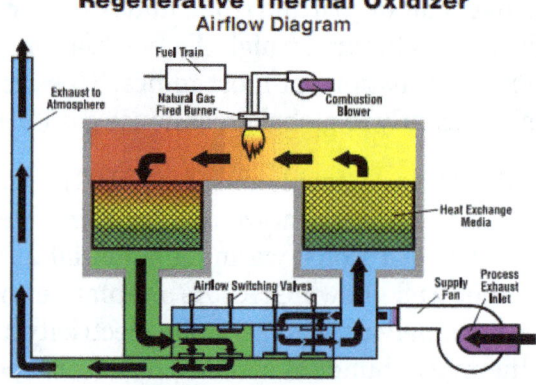

Regenerative Thermal Oxidizer
Airflow Diagram

2. Inlet switch valves direct the pollutant-laden air into one of the energy recovery canisters.

3. The air continues to travel from the valve assembly up through the first heat exchanger canister where heat is absorbed from the ceramic media.

4. The preheated air moves into the combustion chamber and is held at an elevated temperature close to that required for oxidation.

5. The pollutant is destroyed while in the combustion chamber.

The Clean Air Cycle

6. Once out of the combustion chamber, the clean, hot air moves through a second energy recovery canister where the ceramic media absorbs the heat generated during thermal oxidation.

7. Finally, the cooled, clean air is routed into the atmosphere through outlet switch values and ultimately the exhaust stack.

    • *Thermal energy efficiencies range from 85-97%.*

    • *Destruction efficiencies of 99% are typically guaranteed.*

Best Available Technology

'Best available techniques' (BAT) means the available techniques which are the best for preventing or minimising emissions and impacts on the environment. You need to use BAT if your operation is an installation (eg a facility that carries out an industrial process like a refinery, food factory or intensive farm).

'Techniques' include both the technology used and the way your installation is designed, built, maintained, operated and decommissioned.

The European Commission produces best available technique reference documents or BREF notes. They contain 'best available techniques' (BAT) for installations.

For example, there's a BREF for intensive agriculture which contains BAT for housing for pig rearing units and a BREF for the textiles industry which contains BAT for selecting materials for textile manufacture.

The European Commission is updating BREF notes and the updated versions also include 'BAT conclusion documents'. These contain emission limits associated with BAT ('BAT AELs') which must be complied with unless the Environment Agency agrees you've met certain criteria.

Ways to follow BAT

Your permit conditions may tell you what BAT you must use or they may set emission limit values (ELV) or other environmental outcomes, based on BAT.

If your permit says you must follow BAT or 'appropriate measures' to achieve an outcome or ELV, check the guide for your activity for the BAT for that process.

You may have to decide which BAT to use yourself if your permit doesn't tell you which BAT to use.

You may also need to take additional measures to meet the conditions in your permit.

BAT in your Permit Application

When you apply for an environmental permit you must state whether you're going to follow each BAT that applies to your activity, or propose an alternative.

You need to do this in the 'operating techniques' section of the application form.

For BAT that you're proposing to follow, you must explain how you're going to either:

- follow the BAT conclusions and meet the BAT-associated emissions level (for BAT that are contained in BAT conclusions).

- follow the BREF note and the technical guidance for activities that don't have BAT conclusions.

The origin of the "best available technology" is derived from the second half of the 1800s. It has been expressed in various ways, such as "Best practicable means", but has always had basically the same meaning. Today's "Best Available Technology" was coined for the first time in the 1992 OSPAR Convention where and accurately described by:

1. The use of the best available techniques shall emphasise the use of non-waste technology, if available.

2. The term "best available techniques" means the latest stage of development (state of the art) of processes, of facilities or of methods of operation which indicate the practical suitability of a particular measure for limiting discharges, emissions and waste. In determining whether a set of processes, facilities and methods of operation constitute the best available techniques in general or individual cases, special consideration shall be given to:

    (a) comparable processes, facilities or methods of operation which have recently been successfully tried out;

    (b) technological advances and changes in scientific knowledge and understanding;

    (c) the economic feasibility of such techniques;

    (d) time limits for installation in both new and existing plants;

    (e) the nature and volume of the discharges and emissions concerned.

3. It therefore follows that what is "best available techniques" for a particular process will change with time in the light of technological advances, economic and social factors, as well as changes in scientific knowledge and understanding.

4. If the reduction of discharges and emissions resulting from the use of best available techniques does not lead to environmentally acceptable results, additional measures have to be applied.

European Union Directives

Best available techniques not entailing excessive costs (BATNEEC), sometimes referred to as *best available technology*, was introduced in 1984 with Directive 84/360/EEC and applied to air pollution emissions from large industrial installations.

In 1996, Directive 84/360/EEC was superseded by the Integrated pollution prevention and control directive (IPPC), 96/61/EC, which applied the framework concept of *Best Available Techniques* (BAT) to the integrated control of pollution to the three media air, water and soil. The concept is also part of the directive's recast in 2008 (2008/1/EC) and its successor directive, the Industrial Emissions Directive 2010/75/EU published in 2010.

According to article 15(2) of the Industrial Emissions Directive, emission limit values and the equivalent parameters and technical measures in permits shall be based on the best available techniques, without prescribing the use of any technique or specific technology.

The directive includes a definition of best available techniques in article 3(10):

"best available techniques" means the most effective and advanced stage in the development of activities and their methods of operation which indicates the practical suitability of particular techniques for providing the basis for emission limit values and other permit conditions designed to prevent and, where that is not practicable, to reduce emissions and the impact on the environment as a whole:

- "techniques" includes both the technology used and the way in which the installation is designed, built, maintained, operated and decommissioned;

- "available" means those developed on a scale which allows implementation in the relevant industrial sector, under economically and technically viable conditions, taking into consideration the costs and advantages, whether or not the techniques are used or produced inside the Member State in question, as long as they are reasonably accessible to the operator;

- "best" means most effective in achieving a high general level of protection of the environment as a whole.

    BAT for a given industrial sector are described in BAT reference documents (BREFs) as defined in article 3(11) of the Industrial Emissions Directive. BREFs are the result of an exchange of information between European Union Member States, the industries concerned, non-governmental organisations promoting environmental protection and the European Commission pursuant to article 13 of the directive. This exchange of information is often called the Sevilla process because it is steered by the Institute for Prospective Technological Studies of the European Commissions' Joint Research Centre, which is based in Seville. The process is described in detail in Commission Implementing Decision 2012/119/EU. The most important chapter of the BREFs, the BAT conclusions, are published as implementing decisions of the European Commission in the Official Journal of the European Union. According to article 14(3) of the Industrial Emissions Directive, the BAT conclusions shall be the reference for setting permit conditions of large industrial installations.

United States Environmental law

Clean Air Act

The Clean Air Act requires that certain facilities employ Best Available Control Technology to control emissions.

> An emission limitation based on the maximum degree of reduction of each pollutant subject to regulation under this Act emitted from or which results from any major emitting facility, which the permitting authority, on a case-by-case basis, taking into account energy, environmental, and economic impacts and other costs, determines is achievable for such facility through application of production processes and available methods, systems, and techniques, including fuel cleaning, clean fuels, or treatment or innovative fuel combustion techniques for control of each such pollutant.

Clean Water Act

The Clean Water Act (CWA) requires issuance of national industrial wastewater discharge regulations (called "effluent guidelines"), which are based on BAT and several related standards.

> Effluent limitations for categories and classes of point sources, which (i) shall require application of the best available technology economically achievable for such category or class, which will result in reasonable further progress toward the national goal of eliminating the discharge of all pollutants. Factors relating to the assessment of best available technology shall take into account the age of equipment and facilities involved, the process employed, the engineering aspects of the application of various types of control techniques, process changes, the cost of achieving such effluent reduction, non-water quality environmental impact (including energy requirements), and such other factors as the Administrator deems appropriate.

> In the development of the effluent standards, the BAT concept is a "model" technology rather than a specific regulatory requirement. The U.S. Environmental Protection Agency(EPA) identifies a particular model technology for an industry, and then writes a regulatory performance standard based on the model. The performance standard is typically expressed as a numeric effluent limit measured at the discharge point. The industrial facility may use any technology that meets the performance standard.

> A related CWA provision for cooling water intake structures requires standards based on "best technology available."

> The location, design, construction, and capacity of cooling water intake structures reflect the best technology available for minimizing adverse environmental impact.

International Conventions

The concept of BAT is also used in a number of international conventions such as the Minamata Convention on Mercury, the Stockholm Convention on Persistent Organic Pollutants, or the OSPAR Convention for the protection of the marine environment of the North-East Atlantic.

Emissions Trading

Emissions trading is a market-based approach to controlling pollution. By creating tradable pollution permits it attempts to add the profit motive as an incentive for good performance, unlike traditional environmental regulation based solely on the threat of penalties.

Power plants in Europe, such as this coal power station in Gelsenkirchen, Germany, are regulated by an emissions trading scheme

Developed in the 70s and 80s, emissions trading was introduced in the US in 1990 to combat acid rain, but more recently it has grown in prominence as a way of tackling greenhouse gas emissions linked to climate change.

The main form of emissions trading is known as "cap and trade": a cap on emissions is set and then permits are created up to the level of this cap. The companies or other entities covered by the scheme need to hold one permit for every tonne of pollution (CO2e) they emit. Allowing a trade in these permits puts a price on pollution – the cost of emitting one tonne of carbon dioxide is the cost of the permit – and creates flexibility as to how and where pollution is reduced.

The theory is that setting a limit on pollution and allowing the market to decide how to stay within that limit is ideally suited to reducing carbon emissions, which come from almost all forms of economic activity and mix into the atmosphere with global effect. The market should ensure that the emissions cuts happen at the lowest possible cost, and the cap can be lowered year by year in a managed way.

Supporters argue that this is preferable to other forms of pricing, such as carbon taxes, which do not guarantee any particular level of reduction. However, critics often emphasise the degree to which emissions trading has been marred by weak caps, free handouts of permits to the biggest polluters and the purchase of "offsets" – carbon credits bought from outside the cap-and-trade system from carbon reduction projects in the developing world.

Emissions trading is a central element of the Kyoto protocol in the form of the Clean Development Mechanism (CDM) and is the cornerstone policy of the EU, whose Emissions Trading System (ETS) is the largest in the world. The expansion of emissions trading was slowed significantly by the US decision to abandon a proposed national policy, although groups of states have set up regional schemes. A number of countries are considering the adoption of some form of cap and trade, including China and South Korea.

A coal power plant in Germany. Due to emissions trading, coal may become a less competitive fuel than other options.

Pollution is the prime example of a market externality. An externality is an effect of some activity on an entity (such as a person) that is not party to a market transaction related to that activity. Emissions trading is a market-based approach to address pollution. The overall goal of an emissions trading plan is to minimize the cost of meeting a set emissions target.

In an emissions trading system, the government sets an overall limit on emissions, and defines permits (also called allowances), or limited authorizations to emit, up to the level of the overall limit. The government may sell the permits, but in many existing schemes, it gives permits to participants (regulated polluters) equal to each participant's baseline emissions. The baseline is determined by reference to the participant's historical emissions. To demonstrate compliance, a participant must hold permits at least equal to the quantity of pollution it actually emitted during the time period. If every participant complies, the total pollution emitted will be at most equal to the sum of individual limits. Because permits can be bought and sold, a participant can choose either to use its permits exactly (by reducing its own emissions); or to emit less than its permits, and perhaps sell the excess permits; or to emit more than its permits, and buy permits from other participants. In effect, the buyer pays a charge for polluting, while the seller gains a reward for having reduced emissions.

In many schemes, organizations which do not pollute (and therefore have no obligations) may also trade permits and financial derivatives of permits. In some schemes, participants can bank allowances to use in future periods. In some schemes, a proportion of all traded permits must be retired periodically, causing a net reduction in emissions over time. Thus, environmental groups may buy and retire permits, driving up the price of the remaining permits according to the law of demand. In most schemes, permit owners can donate permits to a nonprofit entity and receive a tax deduction. Usually, the government lowers the overall limit over time, with an aim towards a national emissions reduction target.

According to the Environmental Defense Fund, cap-and-trade is the most environmentally and economically sensible approach to controlling greenhouse gas emissions, the primary cause of global warming, because it sets a limit on emissions, and the trading encourages companies to innovate in order to emit less.

"International trade can offer a range of positive and negative incentives to promote international cooperation on climate change (robust evidence, medium agreement). Three issues are key to developing constructive relationships between international trade and climate agreements: how existing trade policies and rules can be modified to be more climate friendly; whether border adjustment measures (BAMs) or other trade measures can be effective in meeting the goals of international climate agreements; whether the UNFCCC, World Trade Organization (WTO), hybrid of the two, or a new institution is the best forum for a trade-and-climate architecture."

Market and Least-cost

Some economists have urged the use of market-based instruments such as emissions trading to address environmental problems instead of prescriptive "command-and-control" regulation. Command and control regulation is criticized for being insensitive to geographical and technological differences, and therefore inefficient.; however, this is not always so, as shown by the WW-II rationing program in the U.S. in which local and regional boards made adjustments for these differences.

After an emissions limit has been set by a government political process, individual companies are free to choose how or whether to reduce their emissions. Failure to report emissions and surrender emission permits is often punishable by a further government regulatory mechanism, such as a fine that increases costs of production. Firms will choose the least-cost way to comply with the pollution regulation, which will lead to reductions where the least expensive solutions exist, while allowing emissions that are more expensive to reduce.

Under an emissions trading system, each regulated polluter has flexibility to use the most cost-effective combination of buying or selling emission permits, reducing its emissions by installing cleaner technology, or reducing its emissions by reducing production. The most cost-effective strategy depends on the polluter's marginal abatement cost and the market price of permits. In theory, a polluter's decisions should lead to an economically efficient allocation of reductions among polluters, and lower compliance costs for individual firms and for the economy overall, compared to command-and-control mechanisms.

Emission Markets

For emissions trading where greenhouse gases are regulated, one emissions permit is considered equivalent to one metric ton of carbon dioxide (CO_2) emissions. Other names for emissions permits are carbon credits, Kyoto units, assigned amount units, and Certified Emission Reduction units (CER). These permits can be sold privately or in the international market at the prevailing market price. These trade and settle internationally, and hence allow permits to be transferred between countries. Each international transfer is validated by the United Nations Framework Convention on Climate Change (UNFCCC). Each transfer of ownership within the European Union is additionally validated by the European Commission.

Emissions trading programmes such as the European Union Emissions Trading System (EU ETS) complement the country-to-country trading stipulated in the Kyoto Protocol by allowing private trading of permits. Under such programmes – which are generally co-ordinated with the national emissions targets provided within the framework of the Kyoto Protocol – a national or international authority allocates permits to individual companies based on established criteria, with a view to meeting national and/or regional Kyoto targets at the lowest overall economic cost.

Trading exchanges have been established to provide a spot market in permits, as well as futures and options market to help discover a market price and maintain liquidity. Carbon prices are normally quoted in euros per tonne of carbon dioxide or its equivalent (CO_2e). Other greenhouse gases can also be traded, but are quoted as standard multiples of carbon dioxide with respect to their global warming potential. These features reduce the quota's financial impact on business, while ensuring that the quotas are met at a national and international level.

Currently, there are six exchanges trading in UNFCCC related carbon credits: the Chicago Climate Exchange (until 2010), European Climate Exchange, NASDAQ OMX Commodities Europe, PowerNext, Commodity Exchange Bratislava and the European Energy Exchange. NASDAQ OMX Commodities Europe listed a contract to trade offsets generated by a CDM carbon project called Certified Emission Reductions. Many companies now engage in emissions abatement, offsetting, and sequestration programs to generate credits that can be sold on one of the exchanges. At least one private electronic market has been established in 2008: CantorCO2e. Carbon credits at Commodity Exchange Bratislava are traded at special platform called Carbon place.

Trading in emission permits is one of the fastest-growing segments in financial services in the City of London with a market estimated to be worth about €30 billion in 2007. Louis Redshaw, head of environmental markets at Barclays Capital, predicts that "carbon will be the world's biggest commodity market, and it could become the world's biggest market overall."

Pollution Markets

An emission license directly confers a right to emit pollutants up to a certain rate. In contrast, a pollution license for a given location confers the right to emit pollutants at a rate which will cause no more than a specified increase at the pollution-level. For concreteness, consider the following model.

- There are n agents each of which emits e_i pollutants.

- There are m locations each of which suffers pollution q_i.

- The pollution is a linear combination of the emissions. The relation between e and q is given by a *diffusion matrix* H, such that: $q = H \cdot e$.

As an example, consider three countries along a river (as in the fair river sharing setting).

- Pollution in the upstream country is determined only by the emission of the upstream country. $q_1 = e_1$.

- Pollution in the middle country is determined by its own emission and by the emission of country 1: $q_2 = e_1 + e_2$.

- Pollution in the downstream country is the sum of all emissions: $q_3 = e_1 + e_2 + e_3$.

So the matrix H in this case is a triangular matrix of ones.

Each pollution-license for location i permits its holder to emit pollutants that will cause at most this level of pollution at location i. Therefore, a polluter that affects water quality at a number of points has to hold a portfolio of licenses covering all relevant monitoring-points. In the above example, if country 2 wants to emit a unit of pollutant, it should purchase two permits: one for location 2 and one for location 3.

Montgomery shows that, while both markets lead to efficient license allocation, the market in pollution-licenses is more widely applicable than the market in emission-licenses.

Comparison with other Methods of Emission Reduction

Cap and trade is the textbook emissions trading program. Other market-based approaches include baseline-and-credit, and pollution tax. They all put a price on pollution, and so provide an economic incentive to reduce pollution beginning with the lowest-cost opportunities. By contrast, in a command-and-control approach, a central authority designates pollution levels each facility is allowed to emit. Cap and trade essentially functions as a tax where the tax rate is variable based on the relative cost of abatement per unit, and the tax base is variable based on the amount of abatement needed.

Baseline and Credit

In a baseline and credit program, polluters can create permits, called credits or offsets, by reducing their emissions below a baseline level, which is often the historical emissions level from a designated past year. Such credits can be bought by polluters that have a regulatory limit.

Pollution Tax

Emissions fees or environmental tax is a surcharge on the pollution created while producing goods and services. For example, a carbon tax is a tax on the carbon content of fossil fuels that aims to discourage their use and thereby reduce carbon dioxide emissions. The two approaches are overlapping sets of policy designs. Both can have a range of scopes, points of regulation, and price schedules. They can be fair or unfair, depending on how the revenue is used. Both have the effect of increasing the price of goods (such as fossil fuels) to consumers. A comprehensive, upstream, auctioned cap-and-trade system is very similar to a comprehensive, upstream carbon tax. Yet, many commentators sharply contrast the two approaches.

The main difference is what is defined and what derived. A tax is a price control, while cap-and-trade method acts is a quantity control instrument. That is, a tax is a unit price for pollution that is set by authorities, and the market determines the quantity emitted; in cap and trade, authorities determine the amount of pollution, and the market determines the price. This difference affects a number of criteria.

Responsiveness to inflation: Cap-and-trade has the advantage that it adjusts to inflation (changes to overall prices) automatically, while emissions fees must be changed by regulators.

Responsiveness to cost changes: It is not clear which approach is better. It is possible to combine the two into a safety valve price: a price set by regulators, at which polluters can buy additional permits beyond the cap.

Responsiveness to recessions: This point is closely related to responsiveness to cost changes, because recessions cause a drop in demand. Under cap and trade, the emissions cost automatically decreases, so a cap-and-trade scheme adds another automatic stabilizer to the economy - in effect, an automatic fiscal stimulus. However, a lower pollution price also results in reduced efforts to reduce pollution. If the government is able to stimulate the economy regardless of the cap-and-trade scheme, an excessively low price causes a missed opportunity to cut emissions faster than planned. Instead, it might be better to have a price floor (a tax). This is especially true when cutting pollution is urgent, as with greenhouse gas emissions. A price floor also provides certainty and stability for investment in emissions reductions: recent experience from the UK shows that nuclear power operators are reluctant to invest on "un-subsidised" terms unless there is a guaranteed price floor for carbon (which the EU emissions trading scheme does not presently provide).

Responsiveness to uncertainty: As with cost changes, in a world of uncertainty, it is not clear whether emissions fees or cap-and-trade systems are more efficient—it depends on how fast the marginal social benefits of reducing pollution fall with the amount of cleanup (e.g., whether inelastic or elastic marginal social benefit schedule).

Other: The magnitude of the tax will depend on how sensitive the supply of emissions is to the price. The permit price of cap-and-trade will depend on the pollutant market. A tax generates government revenue, but full-auctioned emissions permits can do the same. A similar upstream cap-and-trade system could be implemented. An upstream carbon tax might be the simplest to administer. Setting up a complex cap-and-trade arrangement that is comprehensive has high institutional needs.

Command-and-control Regulation

Command and control is a system of regulation that prescribes emission limits and compliance methods for each facility or source. It is the traditional approach to reducing air pollution.

Command-and-control regulations are more rigid than incentive-based approaches such as pollution fees and cap and trade. An example of this is a performance standard which sets an emissions goal for each polluter that is fixed and, therefore, the burden of reducing pollution cannot be shifted to the firms that can achieve it more cheaply. As a result, performance standards are likely to be more costly overall. The additional costs would be passed to end consumers.

Economics of International Emissions Trading

It is possible for a country to reduce emissions using a Command-Control approach, such as regulation, direct and indirect taxes. The cost of that approach differs between countries because the Marginal Abatement Cost Curve (MAC) — the cost of eliminating an additional unit of pollution — differs by country. It might cost China $2 to eliminate a ton of CO_2, but it would probably cost Norway or the U.S. much more. International emissions-trading markets were created precisely to exploit differing MACs.

Example

Emissions trading through *Gains from Trade* can be more beneficial for both the buyer and the seller than a simple emissions capping scheme.

Consider two European countries, such as Germany and Sweden. Each can either reduce all the required amount of emissions by itself or it can choose to buy or sell in the market.

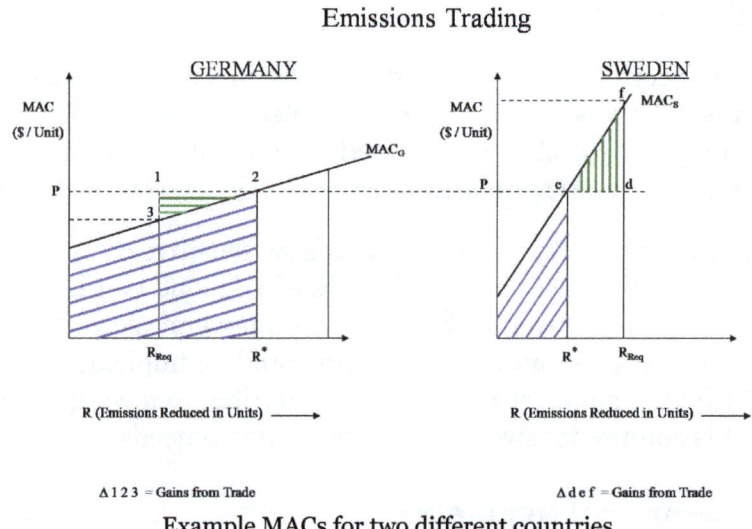

Example MACs for two different countries

Suppose Germany can abate its CO_2 at a much cheaper cost than Sweden, i.e. $MAC_S >$ MAC_G where the MAC curve of Sweden is steeper (higher slope) than that of Germany, and R_{Req} is the total amount of emissions that need to be reduced by a country.

On the left side of the graph is the MAC curve for Germany. R_{Req} is the amount of required reductions for Germany, but at R_{Req} the MAC_G curve has not intersected the market emissions permit price of CO_2 (market permit price = P = λ). Thus, given the market price of CO_2 allowances, Germany has potential to profit if it abates more emissions than required.

On the right side is the MAC curve for Sweden. R_{Req} is the amount of required reductions for Sweden, but the MAC_S curve already intersects the market price of CO_2 permits before R_{Req} has been reached. Thus, given the market price of CO_2 permits, Sweden has potential to make a cost saving if it abates fewer emissions than required internally, and instead abates them elsewhere.

In this example, Sweden would abate emissions until its MAC_S intersects with P (at R*), but this would only reduce a fraction of Sweden's total required abatement.

After that it could buy emissions credits from Germany for the price *P* (per unit). The internal cost of Sweden's own abatement, combined with the permits it buys in the market from Germany, adds up to the total required reductions (R_{Req}) for Sweden. Thus Sweden can make a saving from buying permits in the market (Δ d-e-f). This represents the "Gains from Trade", the amount of additional expense that Sweden would otherwise have to spend if it abated all of its required emissions by itself without trading.

Germany made a profit on its additional emissions abatement, above what was required: it met the regulations by abating all of the emissions that was required of it (R_{Req}). Additionally, Germany sold its surplus permits to Sweden, and was paid P for every unit it abated, while spending less than P. Its total revenue is the area of the graph (R_{Req} 1 2 R*), its total abatement cost is area (R_{Req} 3 2 R*), and so its net benefit from selling emission permits is the area (Δ 1-2-3) i.e. Gains from Trade.

The two R* (on both graphs) represent the efficient allocations that arise from trading.

- Germany: sold (R* - R_{Req}) emission permits to Sweden at a unit price P.

- Sweden bought emission permits from Germany at a unit price P.

If the total cost for reducing a particular amount of emissions in the *Command Control* scenario is called X, then to reduce the same amount of combined pollution in Sweden and Germany, the total abatement cost would be less in the *Emissions Trading* scenario i.e. ($X - \Delta$ 123 - Δ def).

The example above applies not just at the national level, but also between two companies in different countries, or between two subsidiaries within the same company.

Applying the Economic Theory

The nature of the pollutant plays a very important role when policy-makers decide which framework should be used to control pollution. CO_2 acts globally, thus its impact on the environment is generally similar wherever in the globe it is released. So the location of the originator of the emissions does not matter from an environmental standpoint.

The policy framework should be different for regional pollutants (e.g. SO_2 and NO_x, and also mercury) because the impact of these pollutants may differ by location. The same amount of a regional pollutant can exert a very high impact in some locations and a low impact in other locations, so it matters where the pollutant is released. This is known as the *Hot Spot* problem.

A Lagrange framework is commonly used to determine the least cost of achieving an objective, in this case the total reduction in emissions required in a year. In some cases, it is possible to use the Lagrange optimization framework to determine the required reductions for each country (based on their MAC) so that the total cost of reduction is minimized. In such a scenario, the Lagrange multiplier represents the market allowance price (P) of a pollutant, such as the current market price of emission permits in Europe and the USA.

Countries face the permit market price that exists in the market that day, so they are able to make individual decisions that would minimize their costs while at the same time achieving regulatory compliance. This is also another version of the Equi-Marginal Principle, commonly used in economics to choose the most economically efficient decision.

Prices Versus Quantities and the Safety Valve

There has been longstanding debate on the relative merits of *price* versus *quantity* instruments to achieve emission reductions.

An emission cap and permit trading system is a *quantity* instrument because it fixes the overall emission level (quantity) and allows the price to vary. Uncertainty in future supply and demand conditions (market volatility) coupled with a fixed number of pollution permits creates an uncertainty in the future price of pollution permits, and the industry must accordingly bear the cost of adapting to these volatile market conditions. The burden of a volatile market thus lies with the industry rather than the controlling agency, which is generally more efficient. However, under volatile market conditions, the ability of the controlling agency to alter the caps will translate into an ability to pick "winners and losers" and thus presents an opportunity for corruption.

In contrast, an emission tax is a *price* instrument because it fixes the price while the emission level is allowed to vary according to economic activity. A major drawback of an emission tax is that the environmental outcome (e.g. a limit on the amount of emissions) is not guaranteed. On one hand, a tax will remove capital from the industry, suppressing possibly useful economic activity, but conversely, the polluter will not need to hedge as much against future uncertainty since the amount of tax will track with profits. The burden of a volatile market will be borne by the controlling (taxing) agency rather than the industry itself, which is generally less efficient. An advantage is that, given a uniform tax rate and a volatile market, the taxing entity will not be in a position to pick "winners and losers" and the opportunity for corruption will be less.

Assuming no corruption and assuming that the controlling agency and the industry are equally efficient at adapting to volatile market conditions, the best choice depends on the sensitivity of the costs of emission reduction, compared to the sensitivity of the benefits (i.e., climate damage avoided by a reduction) when the level of emission control is varied.

Because there is high uncertainty in the compliance costs of firms, some argue that the optimum choice is the price mechanism. However, the burden of uncertainty cannot be eliminated, and in this case it is shifted to the taxing agency itself.

The overwhelming majority of climate scientists have repeatedly warned of a threshold in atmospheric concentrations of carbon dioxide beyond which a run-away warming effect could take place, with a large possibility of causing irreversible damage.

With such a risk, a quantity instrument may be a better choice because the quantity of emissions may be capped with more certainty. However, this may not be true if this risk exists but cannot be attached to a known level of greenhouse gas (GHG) concentration or a known emission pathway.

A third option, known as a *safety valve*, is a hybrid of the price and quantity instruments. The system is essentially an emission cap and permit trading system but the maximum (or minimum) permit price is capped. Emitters have the choice of either obtaining permits in the marketplace or buying them from the government at a specified trigger price (which could be adjusted over time). The system is sometimes recommended as a way of overcoming the fundamental disadvantages of both systems by giving governments the flexibility to adjust the system as new information comes to light. It can be shown that by setting the trigger price high enough, or the number of permits low enough, the safety valve can be used to mimic either a pure quantity or pure price mechanism.

All three methods are being used as policy instruments to control greenhouse gas emissions: the EU-ETS is a *quantity* system using the cap and trading system to meet targets set by National Allocation Plans; Denmark has a price system using a carbon tax (World Bank, 2010, p. 218), while China uses the CO_2 market price for funding of its Clean Development Mechanism projects, but imposes a *safety valve* of a minimum price per tonne of CO_2.

Carbon Leakage

Carbon leakage is the effect that regulation of emissions in one country/sector has on the emissions in other countries/sectors that are not subject to the same regulation. There is no consensus over the magnitude of long-term carbon leakage.

In the Kyoto Protocol, Annex I countries are subject to caps on emissions, but non-Annex I countries are not. Barker *et al.* (2007) assessed the literature on leakage. The leakage rate is defined as the increase in CO_2 emissions outside the countries taking domestic mitigation action, divided by the reduction in emissions of countries taking domestic mitigation action. Accordingly, a leakage rate greater than 100% means that actions to reduce emissions within countries had the effect of increasing emissions in other countries to a greater extent, i.e., domestic mitigation action had actually led to an increase in global emissions.

Estimates of leakage rates for action under the Kyoto Protocol ranged from 5% to 20% as a result of a loss in price competitiveness, but these leakage rates were considered very uncertain. For energy-intensive industries, the beneficial effects of Annex I actions through technological development were considered possibly substantial. However, this beneficial effect had not been reliably quantified. On the empirical evidence they assessed, Barker *et al.* (2007) concluded that the competitive losses of then-current mitigation actions, e.g., the EU ETS, were not significant.

Under the EU ETS rules Carbon Leakage Exposure Factor is used to determine the volumes of free allocation of emission permits to industrial installations.

Trade

To understand carbon trading, it is important to understand the products that are being traded. The primary product in carbon markets is the trading of GHG emission permits. Under a cap-and-trade system, permits are issued to various entities for the right to emit GHG emissions that meet emission reduction requirement caps.

One of the controversies about carbon mitigation policy is how to "level the playing field" with border adjustments. For example, one component of the American Clean Energy and Security Act (a 2009 bill that did not pass), along with several other energy bills put before US Congress, calls for carbon surcharges on goods imported from countries without cap-and-trade programs. Besides issues of compliance with the General Agreement on Tariffs and Trade, such border adjustments presume that the producing countries bear responsibility for the carbon emissions.

A general perception among developing countries is that discussion of climate change in trade negotiations could lead to "green protectionism" by high-income countries (World Bank, 2010, p. 251). Tariffs on imports ("virtual carbon") consistent with a carbon price of $50 per ton of CO_2 could be significant for developing countries. World Bank (2010) commented that introducing border tariffs could lead to a proliferation of trade measures where the competitive playing field is viewed as being uneven. Tariffs could also be a burden on low-income countries that have contributed very little to the problem of climate change.

Dust Collection System

A dust collection system is an air quality improvement system used in industrial, commercial, and home production shops to improve breathable air quality and safety by removing particulate matter from the air and environment. Dust collection systems work on the basic formula of *capture, convey* and *collect*.

First, the dust must be *captured*. This is accomplished with devices such as capture hoods to catch dust at its source of origin. Many times, the machine producing the dust will have a port to which a duct can be directly attached.

Second, the dust must be *conveyed*. This is done via a ducting system, properly sized and manifolded to maintain a consistent minimum air velocity required to keep the dust in suspension for conveyance to the collection device. A duct of the wrong size can lead to material settling in the duct system and clogging it.

Finally, the dust is *collected*. This is done via a variety of means, depending on the application and the dust being handled. It can be as simple as a basic pass-through filter, a cyclonic separator, or an impingement baffle. It can also be as complex as an electrostatic precipitator, a multistage baghouse, or a chemically treated wet scrubber or stripping tower.

Types of Systems

Smaller dust collection systems use a single-stage vacuum unit to create suction and perform air filtration, where the waste material is drawn into an impeller and deposited into a container such as a bag, barrel, or canister. Air is recirculated into the shop after passing through a filter to trap smaller particulate.

Larger systems utilize a two-stage system, which separates larger particles from fine dust using a pre-collection device, such as a cyclone or baffled canister, before drawing the air through the impeller. Air from these units can then be exhausted outdoors or filtered and recirculated back into the work space.

Dust collection systems are often part of a larger air quality management program that also includes large airborne particle filtration units mounted to the ceiling of shop spaces and mask systems to be worn by workers. Air filtration units are designed to process large volumes of air to remove fine particles (2 to 10 micrometres) suspended in the air. Masks are available in a variety of forms, from simple cotton face masks to elaborate respirators with tanked air — the need for which is determined by the environment in which the worker is operating.

In industry, round or rectangular ducts are used to prevent buildup of dust in processing equipment.

Dangers of Neglect

Proper dust collection and air filtration is important in any work space. Repeated exposure to wood dust can cause chronic bronchitis, emphysema, "flu-like" symptoms, and cancer. Wood dust also frequently contains chemicals and fungi, which can become airborne and lodge deeply in the lungs, causing illness and damage.

Carbon Dioxide Removal

These strategies deal with the root causes of climate change and involve reducing the amount of carbon dioxide already in the atmosphere (different from reducing the amount of carbon dioxide emitted). One method involves capturing carbon dioxide from ambient air, compressing it and storing it in geologic reservoirs (usually porous rock or depleted oil wells). However, at this point in time this strategy is considered too expensive to be cost effective.

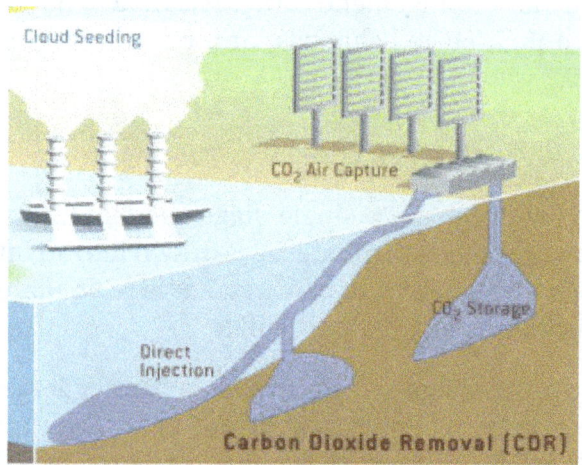

Another strategy, known as Ocean Iron Fertilization, involves adding iron to the ocean in order to increase algal blooms. These blooms consume carbon dioxide as they grow, and when they die, they sink to the ocean floor, effectively sequestering the carbon. However, this method is highly controversial, because it is not known how long the carbon could actually be stored for, or the negative impacts it could have on the ocean ecosystem (2). An approach similar to this involves adding minerals like calcium or magnesium oxide to the ocean. These minerals react with acidic carbon dioxide, and increase the ocean's capacity to hold and store carbon. Another benefit to this method is that it also helps to reduce ocean acidification. However, as with all CDR strategies, there has been little research into long term effects.

CDR strategies are generally not the first ones considered by those who are interested in the use of geoengineering. These strategies are thought of as needing to occur on a very large scale over a long period of time in order to actually be effective. For this reason, they tend to be very expensive and not a viable option at this time.

General

Carbon dioxide removal is different from reducing emissions, as the former produces an outlet of carbon dioxide from Earth's atmosphere, whereas the latter decreases the inlet of carbon dioxide to the atmosphere. Both have the same net effect, but for achieving carbon dioxide concentration levels below present levels, carbon dioxide removal is critical. Also for meeting higher concentration levels, carbon dioxide removal is increasingly considered to be crucial as it provides the only possibility to fill the gap between needed reductions to meet mitigation targets and global emission trends.

In the *OECD Environmental Outlook to 2050* released at the 2011 United Nations Climate Change Conference, the authors commented on the need for negative emissions, stating "Achieving lower concentration targets (450 ppm) depends significantly on the use of BECCS".

A carbon dioxide sink such as a concentrated group of plants or any other primary producer that binds carbon dioxide into biomass, such as within forests and kelp beds, is not carbon negative, as sinks are not permanent. A carbon dioxide sink of this type moves carbon, in the form of carbon dioxide, from the atmosphere or hydrosphere to the biosphere. This process could be undone, for example by wildfires or logging.

Carbon dioxide sinks that store carbon dioxide in the Earth's crust by injecting it into the subsurface, or in the form of insoluble carbonate salts (mineral sequestration), are considered carbon negative. This is because they are removing carbon from the atmosphere and sequestering it indefinitely and presumably for a considerable duration (thousands to millions of years). However, Carbon Capture technology remains, at best, theoretical and is yet to reach more than 33% efficiency. Furthermore, this process could be rapidly undone, for example by earthquakes or mining.

Methods

Bio-energy with Carbon Capture and Storage

Bio-energy with carbon capture and storage, or BECCS, uses biomass to extract carbon dioxide from the atmosphere, and carbon capture and storage technologies to concentrate and permanently store it in deep geological formations.

BECCS is currently (as of October 2012) the only CDR technology deployed at full industrial scale, with 550 000 tonnes CO_2/year in total capacity operating, divided between three different facilities (as of January 2012).

The Imperial College London, the UK Met Office Hadley Centre for Climate Prediction and Research, the Tyndall Centre for Climate Change Research, the Walker Institute for Climate System Research, and the Grantham Institute for Climate Change issued a joint report on carbon dioxide removal technologies as part of the *AVOID: Avoiding dangerous climate change* research program, stating that "Overall, of the technologies studied in this report, BECCS has the greatest maturity and there are no major practical barriers to its introduction into today's energy system. The presence of a primary product will support early deployment."

According to the OECD, "Achieving lower concentration targets (450 ppm) depends significantly on the use of BECCS".

Biochar

Biochar is created by the pyrolysis of biomass, and is under investigation as a method of carbon sequestration. Biochar is a charcoal that is used for agricultural purposes which also aids in carbon sequestration, the capture or hold of carbon. It is created using a process called pyrolysis, which is basically the act of high temperature heating biomass in an environment with low oxygen levels. What remains is a material known as char, similar to charcoal but is made through a sustainable process, thus the use of biomass. Biomass is organic matter produced by living organisms or recently living organisms, most commonly plants or plant based material. The offset of GHG emission, if biochar were to be implemented, would be a maximum of 12%. This equates to about 106 metric tons of CO_2 equivalents. On a medium conservative level, it would be 23% less than that, at 82 metric tons. A study done by the UK Biochar Research Center has stated that, on a conservative level, biochar can store 1 gigaton of carbon per year. With greater effort in marketing and acceptance of biochar, the benefit could be the storage of 5–9 gigatons per year of carbon in biochar soils.

Enhanced Weathering

Enhanced weathering refers to chemical approach to remove carbon dioxide involving land or ocean-based techniques. Examples of land based enhanced weathering techniques are in-situ carbonation of silicates. Ultramafic rock, for example, has the potential to store from hundreds to thousands of years worth of CO_2 emissions according to estimates. Ocean-based techniques involve alkalinity enhancement, such as, grinding, dispersing and dissolving olivine, limestone, silicates, or calcium hydroxide to address ocean acidification and CO_2 sequestration. Enhanced weathering is considered as one of the least expensive of geoengineering options. One example of a research project on the feasibility of enhanced weathering is the CarbFix project in Iceland.

Direct Air Capture (DAC)

Carbon dioxide can be removed from ambient air through chemical processes, sequestered, and stored. Traditional modes of carbon capture such as precombustion and postcombustion CO_2 capture from large point sources can help slow the rate of increase of the atmospheric CO_2 concentration, but only the direct removal of CO_2 from the air, or "direct air capture" (DAC), can actually reduce the global atmospheric CO_2 concentration if combined with long-term storage of CO_2.

A few engineering proposals have been made for removing CO_2 from the atmosphere, but work in this area is still in its infancy. Among the main technologies proposed, three of them stand out: Causticization with alkali and alkali-earth hydroxides, carbonation, and organic–inorganic hybrid sorbents consisting of amines supported in porous adsorbents.

One proposed method is by so-called *artificial trees*. This concept, proposed by climate scientist Wallace S. Broecker and science writer Robert Kunzig, imagines huge numbers of artificial trees around the world to remove ambient CO_2. The technology is now being pioneered by Klaus Lackner, a researcher at the Earth Institute, Columbia

University, whose artificial tree technology can suck up to 1,000 times more CO_2 from the air than real trees can, at a rate of about one ton of carbon per day if the artificial tree is approximately the size of an actual tree. The CO_2 would be captured in a filter and then removed from the filter and stored.

The chemistry used is a variant of that described below, as it is based on sodium hydroxide. However, in a more recent design proposed by Klaus Lackner, the process can be carried out at only 40 °C by using a polymer-based ion exchange resin, which takes advantage of changes in humidity to prompt the release of captured CO_2, instead of using a kiln. This reduces the energy required to operate the process.

Another substance which can be used are Metal-organic frameworks (or MOF's). A special MOF has been made specifically for locking CO_2 by Joeri Denayer.

In 2008, the Discovery Channel covered the work of David Keith, of University of Calgary, who built a tower, 4 feet wide and 20 feet tall (1.2×6.1 meters), with a fan at the bottom that sucks air in, which comes out again at the top. In the process, about half the CO_2 is removed from the air.

This device uses the chemical process described in detail below. The system demonstrated on the Discovery Channel was a 1/90,000th scale test system of the capture section; the reagents are regenerated in a separate facility. The main costs of a full plant will be the cost to build it, and the energy input to regenerate the chemicals and produce a pure stream of CO_2.

To put this into perspective, people in the U.S. emit about 20 tonnes of CO_2 per person annually. In other words, each person in the U.S. would require a tower like the one featured by the Discovery Channel to remove this amount of CO_2 from the air, requiring an annual 2 megawatt-hours of electricity to operate it. By comparison, a refrigerator consumes about 1.2 megawatt-hours annually (2001 figures). But, by combining many small systems such as this into one large system, the construction costs and energy use can be reduced.

It has been proposed that the Solar updraft tower to generate electricity from thermal air currents also be used at the same time for amine gravity scrubbing of CO_2. Some heat would be required to regenerate the amine.

A similar CO_2 scrubber has also been built by Carbon Engineering. Besides simply focusing on capturing the CO_2, the company also puts emphasis on reuse of the CO_2, for example in the production of fuels, which would thus be carbon-neutral.

Direct air capture has been proposed as a way of generating carbon-neutral organic chemicals, by harvesting the atmospheric compounds and then using them in the production and synthesis of polymers and fuels.

The Swiss-based Climeworks built the first industrial scale direct air capturing plant in Hinwil, in the canton of Zurich, Switzerland. It started up in May 2017, and could scrub up to 900 metric tons of CO_2 per year using heat from a local waste incineration plant. The CO2 was then pumped into local greenhouses, where it was used to help grow vegetables like tomatoes and cucumbers.

Climeworks and Reykjavik Energy also started up a small test direct air capturing plant in Hellisheidi, Iceland in October 2017. Using waste heat from a local geothermal power plant, the test plant captured up to 50 tons of CO_2 per year which was bound with water and injected over 700 metres underground into basaltic bedrock. The CO_2 reacted with the basalt to form solid carbonate minerals.

Ocean Fertilization

Ocean fertilization or ocean nourishment is a type of climate engineering based on the purposeful introduction of nutrients to the upper ocean to increase marine food production and to remove carbon dioxide from the atmosphere. A number of techniques, including fertilization by iron, urea and phosphorus have been proposed.

Diesel Particulate Filter

Diesel particulate filters (DPF) are devices that physically capture diesel particulates to prevent their release to the atmosphere. Diesel particulate filter materials have been developed that show impressive filtration efficiencies, in excess of 90%, as well as good mechanical and thermal durability. Diesel particulate filters have become the most effective technology for the control of diesel particulate emissions—including particle mass and numbers—with high efficiencies.

Due to the particle deposition mechanisms in these devices, filters are most effective in controlling the solid fraction of diesel particulates, including elemental carbon (soot) and the related black smoke emission. Filters may have limited effectiveness, or be totally ineffective, in controlling non-solid fractions of PM emissions—SOF and sulfate particulates. To control total PM emissions, DPF systems are likely to incorporate additional functional components targeting the SOF—typically oxidation catalysts—while ultra low sulfur fuels may be required to control sulfate particulates.

The term "diesel particulate trap" is sometimes used as a synonym for "diesel particulate filter", especially in older literature. The term "trap" covers a wider class of particle separation devices. Several particle deposition mechanisms other than filtration are commonly employed in industrial dust separation equipment. Examples include gravity settling, centrifugal separation, or electrostatic trapping. None of these techniques could be adopted to control diesel PM emissions, due to the small particle size and low density of diesel soot.

It may be noted that *particle oxidation catalysts* (POC)—sometimes called *partial filters*—can also capture diesel particulates, but provide a much lower overall efficiency than diesel particulate filters. In their common designs, POCs capture particulates only from a fraction of the flow, whereas the total flow is filtered in diesel particulate filters. In the case of some filter media, however, the distinction may not be very clear and the devices can be classified as either a POC or a (depth) particulate filter.

Collection & Regeneration: Due to the low bulk density of diesel particulates, which is typically below 0.1 g/cm³ (the density depends on the degree of compactness; as an example, a number of 0.056 g/cm³ was reported by Wade [Wade 1981]), diesel particulate filters can quickly accumulate considerable volumes of soot. Several liters of soot per day may be collected from an older generation heavy-duty truck or bus engine. The collected particulates would eventually cause excessively high exhaust gas pressure drop in the filter, which would negatively affect the engine operation. Therefore, diesel particulate filter systems have to provide a way of removing particulates from the filter to restore its soot collection capacity. This removal of particulates, known as the filter *regeneration*, can be performed either continuously, during regular operation of the filter, or periodically, after a pre-determined quantity of soot has been accumulated. In either case, the regeneration of filter systems should be "invisible" to the vehicle driver/operator and should be performed without his intervention. Thermal regeneration of diesel particulate filters is typically employed, where the collected particulates are oxidized—by oxygen and/or nitrogen dioxide—to gaseous products, primarily to carbon dioxide. Thermal regeneration, schematically represented in Figure, is undoubtedly the cleanest and most attractive method of operating diesel particulate filters.

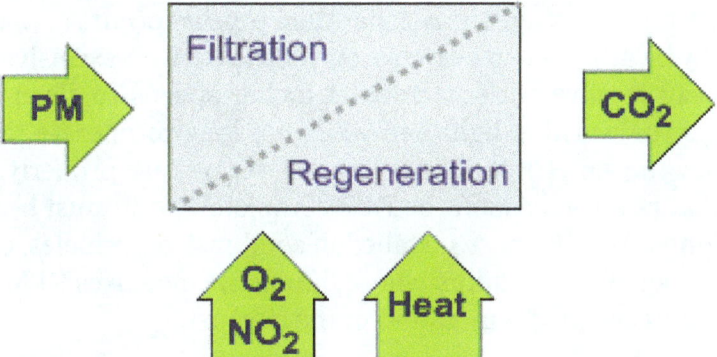

Schematic of particulate filter with thermal regeneration

To ensure that particulates are oxidized at a sufficient rate, the filter must operate at a sufficient temperature. In some filter systems, the source of heat is the exhaust gas stream itself. In this type of filter system, referred to as a *passive filter*, the filter regenerates continuously during the regular operation of the engine. Passive filters usually incorporate some form of a catalyst, which lowers the soot oxidation temperature to a level that can be reached by exhaust gases during the operation of the vehicle.

Another approach which may be needed to facilitate reliable regeneration involves a number of active strategies for increasing the filter temperature (engine management, fuel combustion in the exhaust system, electric heaters, etc.). Regeneration of such devices, known as *active filters*, is usually performed periodically, as determined by the vehicle's control system. The classification of particulate filter systems is presented in more detail in the paper on filter systems.

An alternative strategy involves the use of disposable filter cartridges, which are replaced with new units once filled with soot. Particulate filters of this kind are used in some occupational health environments. Such maintenance intensive filter systems are clearly not acceptable in highway vehicle applications.

Mode of Action

Wall-flow diesel particulate filters usually remove 85% or more of the soot, and under certain conditions can attain soot removal efficiencies approaching 100%. Some filters are single-use, intended for disposal and replacement once full of accumulated ash. Others are designed to burn off the accumulated particulate either passively through the use of a catalyst or by active means such as a fuel burner which heats the filter to soot combustion temperatures. This is accomplished by engine programming to run (when the filter is full) in a manner that elevates exhaust temperature, in conjunction with an extra fuel injector in the exhaust stream that injects fuel to react with a catalyst element to burn off accumulated soot in the DPF filter, or through other methods. This is known as "filter regeneration". Cleaning is also required as part of periodic maintenance, and it must be done carefully to avoid damaging the filter. Failure of fuel injectors or turbochargers resulting in contamination of the filter with raw diesel or engine oil can also necessitate cleaning. The regeneration process occurs at road speeds higher than can generally be attained on city streets; vehicles driven exclusively at low speeds in urban traffic can require periodic trips at higher speeds to clean out the DPF. If the driver ignores the warning light and waits too long to operate the vehicle above 40 miles per hour (64 km/h), the DPF may not regenerate properly, and continued operation past that point may spoil the DPF completely so it must be replaced. Some newer diesel engines, namely those installed in combination vehicles, can also perform what is called a Parked Regeneration, where the engine increases RPM to around 1400 while parked, to increase the temperature of the exhaust.

Diesel engines produce a variety of particles during combustion of the fuel/air mix due to incomplete combustion. The composition of the particles varies widely dependent upon engine type, age, and the emissions specification that the engine was designed to meet. Two-stroke diesel engines produce more particulate per unit of power than do four-strokediesel engines, as they burn the fuel-air mix less completely.

Diesel particulate matter resulting from the incomplete combustion of diesel fuel produces soot (black carbon) particles. These particles include tiny nanoparticles—

smaller than a thousandth of a millimeter (one micron). Soot and other particles from diesel engines worsen the particulate matter pollution in the air and are harmful to health.

New particulate filters can capture from 30% to greater than 95% of the harmful soot. With an optimal diesel particulate filter (DPF), soot emissions may be decreased to 0.001 g / km or less.

The quality of the fuel also influences the formation of these particles. For example, a high sulfur content diesel produces more particles. Lower sulfur fuel produces fewer particles, and allows use of particulate filters. The injection pressure of diesel also influences the formation of fine particles.

DPF and NOx emissions strategies greatly increased fuel consumption in 2007 model year diesel engines, the addition of DEF fluid has reduced fuel consumption, but fuel consumption is still higher than in pre-emissions engines.

Variants of DPFs

Cordierite Diesel Particulate Filter on GM 7.8 Isuzu

Unlike a catalytic converter which is a flow-through device, a DPF retains bigger exhaust gas particles by forcing the gas to flow through the filter; however, the DPF does not retain small particles and maintenance-free DPFs break larger particles into smaller ones. There are a variety of diesel particulate filter technologies on the market. Each is designed around similar requirements:

1. Fine filtration

2. Minimum pressure drop

3. Low cost

4. Mass production suitability

5. Product durability

Cordierite Wall Flow Filters

The most common filter is made of cordierite (a ceramic material that is also used as catalytic converter supports (cores)). Cordierite filters provide excellent filtration efficiency, are relatively inexpensive, and have thermal properties that make packaging them for installation in the vehicle simple. The major drawback is that cordierite has a relatively low melting point (about 1200 °C) and cordierite substrates have been known to melt during filter regeneration. This is mostly an issue if the filter has become loaded more heavily than usual, and is more of an issue with passive systems than with active systems, unless there is a system break down.

Cordierite filter cores look like catalytic converter cores that have had alternate channels plugged - the plugs force the exhaust gas flow through the wall and the particulate collects on the inlet face.

Silicon Carbide Wall Flow Filters

The second most popular filter material is silicon carbide, or SiC. It has a higher (2700 °C) melting point than cordierite, however it is not as stable thermally, making packaging an issue. Small SiC cores are made of single pieces, while larger cores are made in segments, which are separated by a special cement so that heat expansion of the core will be taken up by the cement, and not the package. SiC cores are usually more expensive than cordierite cores, however they are manufactured in similar sizes, and one can often be used to replace the other. Silicon carbide filter cores also look like catalytic converter cores that have had alternate channels plugged - again the plugs force the exhaust gas flow through the wall and the particulate collects on the inlet face.

The characteristics of the wall flow diesel Particulate filter substrate are as follows: broad band filtration (the diameters of the filtered particles are 0.2-150 µm); high filtration efficiency (can be up to 95%); high refractory; high mechanical properties; high boiling point.

Ceramic Fiber Filters

Fibrous ceramic filters are made from several different types of ceramic fibers that are mixed together to form a porous media. This media can be formed into almost any shape and can be customized to suit various applications. The porosity can be controlled in order to produce high flow, lower efficiency or high efficiency lower volume filtration. Fibrous filters have an advantage over wall flow design of producing lower back pressure. Ceramic wall-flow filters remove carbon particulates almost completely, including fine particulates less than 100 nanometers (nm) diameter with an efficiency of greater than 95% in mass and greater than 99% in number of particles over a wide range of engine operating conditions. Since the continuous flow of soot into the filter would eventually block it, it is necessary to 'regenerate' the filtration properties of the filter by burning-off the collected particulate on a regular basis. Soot particulates burn-off forms water and CO_2 in small quantity amounting to less than 0.05% of the CO_2 emitted by the engine.

Metal Fiber flow-through Filters

Some cores are made from metal fibers - generally the fibers are "woven" into a monolith. Such cores have the advantage that an electrical current can be passed through the monolith to heat the core for regeneration purposes, allowing the filter to regenerate at low exhaust temperatures and/or low exhaust flow rates. Metal fiber cores tend to be more expensive than cordierite or silicon carbide cores, and generally not interchangeable with them because of the electrical requirement.

Paper

Disposable paper cores are used in certain specialty applications, without a regeneration strategy. Coal mines are common users — the exhaust gas is usually first passed through a water trap to cool it, and then through the filter. Paper filters are also used when a diesel machine must be used indoors for short periods of time, such as on a forklift being used to install equipment inside a store.

Partial Filters

There are a variety of devices that produce over 50% particulate matter filtration, but less than 85%. Partial filters come in a variety of materials. The only commonality between them is that they produce more back pressure than a catalytic converter, and less than a diesel particulate filter. Partial filter technology is popular for retrofit.

Maintenance

Filters require more maintenance than catalytic converters. Ash, a byproduct of oil consumption from normal engine operation, builds up in the filter as it cannot be converted into a gas and pass through the walls of the filter. This increases the pressure before the filter. Warnings are given to the driver before filter restriction causes an issue with drive-ability or damage to the engine or filter develop. Regular filter maintenance is a necessity.

DPF filters go through a regeneration process which removes this soot and lowers the filter pressure. There are three types of regeneration: active, passive, and forced. Active regeneration happens while the vehicle is not in use and takes 10 minutes on average to complete. Passive regeneration takes place while driving using the heat of the exhaust. This works well for vehicles that drive longer distances with few stops compared to those that perform short trips with many starts and stops. If the filter develops too much pressure then the last type of regeneration must be used - a forced regeneration. This involves a garage using a computer program to run the car, initiating a regeneration of the DPF manually.

There are three ways to clean your truck by using three regenerating cycle types. If using the active regeneration, the process ends in about 10 minutes. The long distances truck drivers are well familiarized with the passive regeneration when the cycle once started can only be ended in a mechanic shop. The third regeneration cycle supposes a forced process and occurs when the car is run after a computer program.

Safety

In 2011, Ford recalled 37,400 F-Series trucks with diesel engines after fuel and oil leaks caused fires in the diesel particulate filters of the trucks. No injuries occurred before the recall, though one grass fire was started. A similar recall was issued for 2005-2007 Jaguar S-Type and XJ diesels, where large amounts of soot became trapped in the DPF. In affected vehicles, smoke and fire emanated from the vehicle underside, accompanied by flames from the rear of the exhaust. The heat from the fire could cause heating through the transmission tunnel to the interior, melting interior components and potentially causing interior fires.

Flue-gas Desulfurization

FGD is a set of technologies used to remove sulphurdioxide (SO2) from exhaust Øue gases of fossil-fuel power plants, as well as from the emissions of other SOx emitting processes. Common methods used in it are wet scrubbing method, Wet and Dry lime scrubbing method, Spray-dry scrubbing method, SNOX method, Dry sorbent injection method, etc.

For a typical coal-×red power station, FGD system may remove 90% or more of the SO2 in the flue gases. SO2 emissions are a primary contributor to acid rain and have been regulated by every industrialized nation in the world.

Flue Gases

Flue Gases is mixture of gases produced by combustion of fuel and other materials in power stations and various industrial plants and released via flue (ducts) in atmosphere. It largely contains oxides of nitrogen derived from combustion of air, sulphur oxides, carbon dioxide, carbon monoxide, water vapour, excess oxygen, particulate matter like soot.

FGD Chemistry

Basic Principles

Most FGD systems employ two stages: one for fly ash removal and the other for SO_2 removal. Attempts have been made to remove both the fly ash and SO_2 in one scrubbing vessel. However, these systems experienced severe maintenance problems and low removal efficiency. In wet scrubbing systems, the flue gas normally passes first through a fly ash removal device, either an electrostatic precipitator or a baghouse, and then into the SO_2-absorber. However, in dry injection or spray drying operations, the SO_2 is first reacted with the lime, and then the flue gas passes through a particulate control device.

Another important design consideration associated with wet FGD systems is that the flue gas exiting the absorber is saturated with water and still contains some SO_2. These gases are highly corrosive to any downstream equipment such as fans, ducts, and stacks. Two methods that may minimize corrosion are: (1) reheating the gases to above their dew point, or (2) using materials of construction and designs that allow equipment to withstand the corrosive conditions. Both alternatives are expensive. Engineers determine which method to use on a site-by-site basis.

Scrubbing with an alkali solid or solution

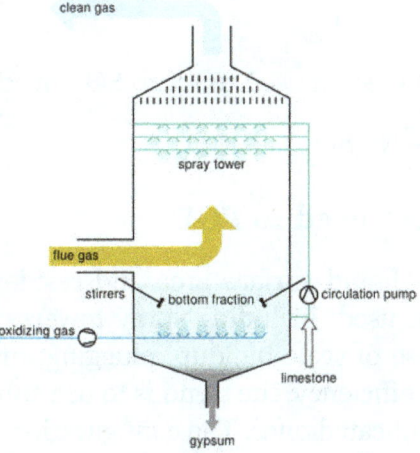

Schematic design of the absorber of an FGD

SO_2 is an acid gas, and, therefore, the typical sorbent slurries or other materials used to remove the SO_2 from the flue gases are alkaline. The reaction taking place in wet scrubbing using a $CaCO_3$ (limestone) slurry produces calcium sulfite ($CaSO_3$) and may be expressed in the simplified dry form as:

$$CaCO_{3(s)} + SO_{2(g)} \rightarrow CaSO_{3(s)} + CO_{2(g)}$$

When wet scrubbing with a $Ca(OH)_2$ (hydrated lime) slurry, the reaction also produces $CaSO_3$ (calcium sulfite) and may be expressed in the simplified dry form as:

$$Ca(OH)_{2(s)} + SO_{2(g)} \rightarrow CaSO_{3(s)} + H_2O_{(l)}$$

When wet scrubbing with a $Mg(OH)_2$ (magnesium hydroxide) slurry, the reaction produces $MgSO_3$ (magnesium sulfite) and may be expressed in the simplified dry form as:

$$Mg(OH)_{2(s)} + SO_{2(g)} \rightarrow MgSO_{3(s)} + H_2O_{(l)}$$

To partially offset the cost of the FGD installation, some designs, particularly dry sorbent injection systems, further oxidize the $CaSO_3$ (calcium sulfite) to produce marketable $CaSO_4\text{-}2H_2O$ (gypsum) that can be of high enough quality to use in wallboard and other products. The process by which this synthetic gypsum is created is also known as forced oxidation:

$$CaSO_{3(aq)} + 2H_2O_{(l)} + \tfrac{1}{2}O_{2(g)} \rightarrow CaSO_4 \cdot 2H_2O_{(s)}$$

A natural alkaline usable to absorb SO_2 is seawater. The SO_2 is absorbed in the water, and when oxygen is added reacts to form sulfate ions $SO_4\text{-}$ and free H^+. The surplus of H^+ is offset by the carbonates in seawater pushing the carbonate equilibrium to release CO_2 gas:

$$SO_{2(g)} + H_2O_{(l)} + \tfrac{1}{2}O_{2(g)} \rightarrow SO_4^{2-}{}_{(aq)} + 2H^+$$

$$HCO_3^- + H^+ \rightarrow H_2O_{(l)} + CO_{2(g)}$$

In industry caustic (NaOH) is often used to scrub SO_2, producing sodium sulfite:

$$2NaOH_{(aq)} + SO_{2(g)} \rightarrow Na_2SO_{3(aq)} + H_2O_{(l)}$$

Types of Wet Scrubbers used in FGD

To promote maximum gas–liquid surface area and residence time, a number of wet scrubber designs have been used, including spray towers, venturis, plate towers, and mobile packed beds. Because of scale buildup, plugging, or erosion, which affect FGD dependability and absorber efficiency, the trend is to use simple scrubbers such as spray towers instead of more complicated ones. The configuration of the tower may be vertical or horizontal, and flue gas can flow cocurrently, countercurrently, or crosscurrently with respect to the liquid. The chief drawback of spray towers is that they require a higher liquid-to-gas ratio requirement for equivalent SO_2 removal than other absorber designs.

FGD scrubbers produce a scaling wastewater that requires treatment to meet discharge regulations. However, technological advancements in ion exchange membranes and el ectrodialysis systems has enabled high-efficiency treatment of FGD wastewater to meet recent EPA discharge limits. The treatment approach is similar for other highly scaling industrial wastewaters.

Venturi-rod Scrubbers

A venturi scrubber is a converging/diverging section of duct. The converging section accelerates the gas stream to high velocity. When the liquid stream is injected at the throat, which is the point of maximum velocity, the turbulence caused by the high gas velocity atomizes the liquid into small droplets, which creates the surface area necessary for mass transfer to take place. The higher the pressure drop in the venturi, the smaller the droplets and the higher the surface area. The penalty is in power consumption.

For simultaneous removal of SO_2 and fly ash, venturi scrubbers can be used. In fact, many of the industrial sodium-based throwaway systems are venturi scrubbers originally designed to remove particulate matter. These units were slightly modified to inject a sodium-based scrubbing liquor. Although removal of both particles and SO_2 in one vessel can be economic, the problems of high pressure drops and finding a scrubbing medium to remove heavy loadings of fly ash must be considered. However, in cases where the particle concentration is low, such as from oil-fired units, it can be more effective to remove particulate and SO_2 simultaneously.

Packed Bed Scrubbers

A packed scrubber consists of a tower with packing material inside. This packing material can be in the shape of saddles, rings, or some highly specialized shapes designed to maximize the contact area between the dirty gas and liquid. Packed towers typically operate at much lower pressure drops than venturi scrubbers and are therefore cheaper to operate. They also typically offer higher SO_2 removal efficiency. The drawback is that they have a greater tendency to plug up if particles are present in excess in the exhaust air stream.

Spray Towers

A spray tower is the simplest type of scrubber. It consists of a tower with spray nozzles, which generate the droplets for surface contact. Spray towers are typically used when circulating a slurry. The high speed of a venturi would cause erosion problems, while a packed tower would plug up if it tried to circulate a slurry.

Counter-current packed towers are infrequently used because they have a tendency to become plugged by collected particles or to scale when lime or limestone scrubbing slurries are used.

Scrubbing Reagent

Alkaline sorbents are used for scrubbing flue gases to remove SO_2. Depending on the application, the two most important are lime and sodium hydroxide (also known as caustic soda). Lime is typically used on large coal- or oil-fired boilers as found in power plants, as it is very much less expensive than caustic soda. The problem is that it results in a slurry being circulated through the scrubber instead of a solution. This makes it harder on the equipment. A spray tower is typically used for this application. The use of lime results in a slurry of calcium sulfite ($CaSO_3$) that must be disposed of. Fortunately, calcium sulfite can be oxidized to produce by-product gypsum ($CaSO_4 \cdot 2H_2O$) which is marketable for use in the building products industry.

Caustic soda is limited to smaller combustion units because it is more expensive than lime, but it has the advantage that it forms a solution rather than a slurry. This makes it easier to operate. It produces a "spent caustic" solution of sodium sulfite/bisulfite (depending on the pH), or sodium sulfate that must be disposed of. This is not a problem in a kraft pulp mill for example, where this can be a source of makeup chemicals to the recovery cycle.

Scrubbing with Sodium Sulfite Solution

It is possible to scrub sulfur dioxide by using a cold solution of sodium sulfite; this forms a sodium hydrogen sulfite solution. By heating this solution it is possible to reverse the reaction to form sulfur dioxide and the sodium sulfite solution. Since the sodium sulfite solution is not consumed, it is called a regenerative treatment. The application of this reaction is also known as the Wellman–Lord process.

In some ways this can be thought of as being similar to the reversible liquid–liquid extraction of an inert gas such as xenon or radon (or some other solute which does not undergo a chemical change during the extraction) from water to another phase. While a chemical change does occur during the extraction of the sulfur dioxide from the gas mixture, it is the case that the extraction equilibrium is shifted by changing the temperature rather than by the use of a chemical reagent.

Gas Phase Oxidation followed by Reaction with Ammonia

A new, emerging flue gas desulfurization technology has been described by the IAEA. It is a radiation technology where an intense beam of electrons is fired into the flue gas at the same time as ammonia is added to the gas. The Chendu power plant in China started up such a flue gas desulfurization unit on a 100 MW scale in 1998. The Pomorzany power plant in Poland also started up a similar sized unit in 2003 and that plant removes both sulfur and nitrogen oxides. Both plants are reported to be operating successfully. However, the accelerator design principles and manufacturing quality need further improvement for continuous operation in industrial conditions.

No radioactivity is required or created in the process. The electron beam is generated by a device similar to the electron gun in a TV set. This device is called an accelerator. This is an example of a radiation chemistry process where the physical effects of radiation are used to process a substance.

The action of the electron beam is to promote the oxidation of sulfur dioxide to sulfur(VI) compounds. The ammonia reacts with the sulfur compounds thus formed to produce ammonium sulfate, which can be used as a nitrogenous fertilizer. In addition, it can be used to lower the nitrogen oxide content of the flue gas. This method has attained industrial plant scale.

Dry Lime Scrubbing

In dry scrubbing, lime is injected directly into flue gas to remove SO_2 and HCl. There are two major dry processes: "dry injection" systems inject dry hydrated lime into the flue gas duct and "spray dryers" inject an atomized lime slurry into a separate vessel. A spray dryer is typically shaped like a silo, with a cylindrical top and a cone bottom. Hot flue gas flows into the top. Lime slurry is sprayed through an atomizer (*e.g.*, nozzles) into the cylinder near the top, where it absorbs SO_2 and HCl. The water in the lime slurry is then evaporated by the hot gas. The scrubbed flue gas flows from the bottom of the cylindrical section through a horizontal duct. A portion of the dried, unreacted lime and its reaction products fall to the bottom of the cone and are removed. The flue gas then flows to a particulate control device (*e.g.*, a baghouse) to remove the remainder of the lime and reaction products. Both dry injection and spray dryers yield a dry final product, collected in particulate control devices. At electric generating plants, dry scrubbing is used primarily for low-sulfur fuels. At municipal waste-to-energy plants, dry scrubbing is used for removal of SO_2 and HCl. Dry scrubbing is used at other industrial facilities for HCl control. Dry scrubbing methods have improved significantly in recent years, resulting in excellent removal efficiencies.

Wet Lime Scrubbing

In wet lime scrubbing, lime is added to water and the resulting slurry is sprayed into a flue gas scrubber. In a typical system, the gas to be cleaned enters the bottom of a cylinder-like tower and flows upward through a shower of lime slurry. The sulfur dioxide is absorbed into the spray and then precipitated as wet calcium sulfite. The sulfite can be converted to gypsum, a salable by-product. Wet scrubbing treats high-sulfur fuels and some low-sulfur fuels where high-efficiency sulfur dioxide removal is required. Wet scrubbing primarily uses magnesium-enhanced lime (containing 3-8% magnesium oxide) because it provides high alkalinity to increase SO_2 removal capacity and reduce scaling potential.

Comparing Lime and Limestone SO_2 Wet Scrubbing Processes

More than ninety percent of U.S. flue gas desulfurization (FGD) system capacity uses lime or limestone. This trend will likely continue into the next phase of federally mandated SO_2 reduction from coal burning power plants. In 2003, the National Lime Association sponsored a study by Sargent and Lundy to compare the costs of leading lime and limestone-based FGD processes utilized by power generating plants in the United States. The study included developing conceptual designs with capital and O&M

cost requirements using up-to-date performance criteria for the processes. The results of the study are summarized in two reports: Wet FGD Technology Evaluation and Dry FGD Technology Evaluation. The reports present the competitive position of wet and dry limestone and lime-based processes relative to reagent cost, auxiliary power cost, coal sulfur content, dispatch, capital cost, and by-product production (gypsum and SO_3 aerosol mitigation chemicals).

References

- Dryzek, John S.; Norgaard, Richard B.; Schlosberg, David (2011). The Oxford Handbook of Climate Change and Society. Oxford University Press. p. 154. ISBN 9780199683420

- "Thermal Oxidizer". U.S. EPA Technology Transfer Network Clearinghouse for Inventories & Emissions Factors. U.S. Environmental Protection Agency. Retrieved 4 April 2015

- Robert W., Hahn (November 2011). "The effect of Allowance Allocations on Cap-and-Trade System Performance" (PDF). Journal of Law and Economics. 54 (4). Retrieved November 22,2014

- Burton, E. S.; Sanjour, William (1970). "A Simulation Approach to Air Pollution Abatement Program Planning". Socio-Economic Planning Science. 4: 147–150. doi:10.1016/0038-0121(70)90036-4

- Burney, Nelson E. (2010). Carbon Tax and Cap-and-trade Tools : Market-based Approaches for Controlling Greenhouse Gases. New York: Nova Science Publishers, Inc. ISBN 9781608761371

- "Catalytic Oxidizer". U.S. EPA Technology Transfer Network Clearinghouse for Inventories & Emissions Factors. U.S. Environmental Protection Agency. Retrieved 4 April 2015

- Voss, Jan-Peter (2007). "Innovation processes in governance: the development of emissions trading as a new policy instrument". Science and Public Policy. 34: 329–343. doi:10.3152/030234207x228584

- Tiwari, Gopal Nath; Agrawal, Basant (2010). Building integrated photovoltaic thermal systems : for sustainable developments. Cambridge: Royal Society of Chemistry. ISBN 1-84973-090-3

- "SWANA 2012 Excellence Award Application "Landfill Gas Control" Seneca Landfill, Inc" (PDF). Retrieved 5 April 2015

- Dales, John H (1968). "Land, Water, and Ownership". The Canadian Journal of Economics. 1 (4): 791–804. JSTOR 133706

- Nikulshina, V.; Ayesa, N.; Gálvez, M. E.; Stainfeld, A. (2016). "Feasibility of Na–Based Thermochemical Cycles for the Capture of CO_2 from air. Thermodynamic and Thermogravimetric Analyses". Chem. Eng. J. 140 (1–3): 62–70. doi:10.1016/j.cej.2007.09.007

- Vincent D. Blondel: Recent Advances in Learning and Control, p. 233, Springer Science & Business Media, 2008, ISBN 9781848001541

- Sinden, Amy (2014). "Cost-Benefit Analysis, Ben Franklin, and the Supreme Court" (PDF). UC Irvine Law Review. University of California, Irvine School of Law. Retrieved 2016-07-04

- Carlson, Curtis; Burtraw, Dallas; Cropper, Maureen; Palmer, Karen L. (2000). "Sulfur dioxide control by electric utilities: What are the gains from trade?". Journal of Political Economy. 108: 1292–1326. doi:10.1086/317681

Chapter 6

Scrubbers: Air Pollution Control Devices

Scrubber systems are pollution control devices that remove gases and particulates from industrial exhaust streams. Science and technology have undergone rapid developments in the past decade, which has resulted in the discovery of significant tools and techniques such as wet scrubbers, carbon dioxide scrubbers and spray towers, which have been extensively detailed in this chapter.

Scrubber

A chemical scrubber.

A scrubber or scrubber system is a system that is used to remove harmful materials from industrial exhaust gases before they are released into the environment. There are two main ways to scrub pollutants out of exhaust, and they are:

- Wet Scrubbing: The removal of harmful components of exhausted flue gases by spraying a liquid substance through the gas.

- Dry Scrubbing: The removal of harmful components of exhausted flue gases by introducing a solid substance to the gas - generally in powdered form.

Both of these methods work similarly and perform the same process of removing pollutants. The main difference is the materials they use to filter the gases. By removing acidic gases from the exhaust before it is released into the sky, scrubbers help prevent the formation of acid rain.

Use

Scrubbing, sometimes referred to as flue gas desulfurization is the most effect sulfur-removal technique that is in widespread use. Removing the sulfur oxides is fairly simple, the flue gases pass through a spray of water in a wet scrubber that contains a variety of chemicals. Generally speaking, the main chemical is calcium carbonate. If a dry scrubber is used, the flue gas comes into contact with pulverized limestone - which is mainly calcium carbonate. The chemical reaction between the calcium carbonate and the sulfur dioxide yield calcium sulfite. This calcium sulfite either falls out of the gas stream or is removed with other particulates.

Scrubbers are very effective, removing about 98% of sulfur from flue gases, but they are very expensive to maintain and install. Additionally, it is energy intensive as the flue gas must be reheated after coming into contact with water vapour in the wet scrubber for the gas to be buoyant enough to exit through the smokestacks.

Environmental Impacts

The use of scrubbers to clean flue gases before they leave the smokestacks has a drastic, beneficial impact on the environment. By collecting particulate matter and acidic gases, the amount of different pollutants that can exit the plant and be introduced into the environment is dramatically reduced. This increases air quality and lowers the health risks for people who could come into contact with the different pollutants.

Although there are many positive side-effects of using scrubbers, there are still waste products from the scrubbing process whether wet or dry scrubbing is used. These by-products must be disposed of safely since they can rarely be reused because of their chemical content. This is one reason that dry scrubbing has become more common, as the sheer volume of the waste products is less significant than the waste from a wet scrubbing operation.

Synonyms, abbreviations and/or process names

- Scrubber
- Absorption

Removed components

- Gaseous components
- Dust (certain types of application)
- Odour (certain types of application)

Diagram

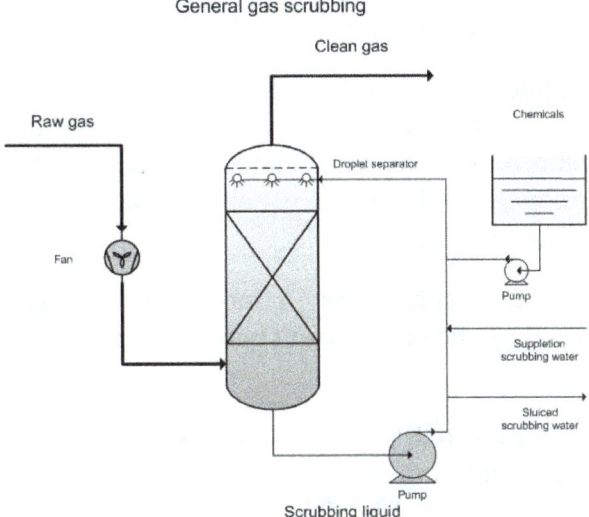

General gas scrubbing

Process Description

A scrubber is a waste gas treatment installation in which a gas stream is brought into intensive contact with a liquid, with the aim of allowing certain gaseous components to pass from the gas to the liquid. Scrubbers can be employed as an emission-limiting technique for many gaseous emissions. Scrubbing is also referred to as absorption.

During scrubbing there is a transfer of components from the gas phase to the liquid phase. The level of gaseous components that can pass to the liquid phase is determined by the ability of these components to dissolve in the liquid. Henry's Law is applicable to the solubility of gases in liquids, for low concentrations and components with a partial pressure < 1atm:

$$p = H.x$$

p = partial pressure (Pa)

x = mole fraction

H = Henry constant (Pa)

This allows one to calculate the maximum concentration of a particular component in the washing water, for the required end concentration. This also gives an indication about water usage under those circumstances.

The equilibrium concentration in the vapour phase, which corresponds to a certain concentration in the liquid phase, is determined by the temperature – the higher the temperature in the liquid phase, the higher the equilibrium concentration in the vapour phase. Thus a reduction in temperature has a favourable effect on the yield.

It is possible to increase the load by adding chemicals to the washing liquid, which help to convert absorbed components. Thus adding chemicals that react with the absorbed gases has a positive effect on the absorption yield.

Besides water (wet scrubbers), organic liquids are also used as absorption mediums. In many cases chemicals or micro-organisms are added to the scrubbing liquid to convert or neutralise gases that are dissolved in the liquid (conditioned scrubbers). As a result of this conversion, the concentration in the water is reduced, which in-turn allows more gas to dissolve (according to Henry's Law).

The concentration of polluted substances in out-going gas streams can never become lower than that permitted by the equilibrium between the gas phase and the scrubbing liquid.

In practice, a scrubber consists of three parts: An absorption section, a droplet collector and a recirculation tank with pump.

Design data:

The liquid-gas ratio (L/G) in a scrubber is the relationship between the scrubbing liquid flow rate and the gas stream flow rate. For dimensioning purposes, and to evaluate the workings of a scrubber, it is important to know how much liquid is required per m³ to realise the required residual emission. The L/G ratio is not only determined by the required residual emission, but is also partly determined by the concentration of the to-be-removed substance(s) in the gas stream and the in and out-going liquid streams. The L/G ratio in a particular situation is thus determined by the selected scrubbing system, the properties of the to-be-cleaned gas, the scrubbing liquid and the to-be-removed component(s) and the requirements set for residual emissions.

Variants

Scrubbers can be distinguished in terms of the flow direction of the gas in relation to the liquid. A distinction is made between counter-flow, co-current and cross-flow scrubbers.

In *counter-flow scrubbing* the scrubbing liquid and the to-be-cleaned gas flow in opposite directions. The main advantage of counter-flow scrubbing is that the cleaner the gas becomes, the lower the pollutant concentration in the scrubbing liquid becomes - whereby the driving force is maintained throughout the column. This type of scrubber is, for example, particularly suited to irregular and peak emissions. The counter-flow set-up allows high concentration peaks to be better dealt with.

In *co-current scrubbers*, the gas and liquid stream move in the same direction. They are less effective than counter-flow scrubbers. However, the advantage they offer is that they are suited to high gas and liquid loads. Co-current scrubbers have a more compact construction and are normally considered when limited space is available and a lower yield is acceptable. Further, they are effective as an initial scrubbing stage for a counter-flow scrubber, for example, when the gas flow needs to be cooled or partly separated.

In *cross-current scrubbers*, the gas and the liquid move across one another. For vapour-like components, the liquid will normally flow in a downward direction and gases will flow horizontally. In dust scrubbing, the sprayers will be horizontal to the gas flow. This type of scrubber is more compact than a counter-current scrubber, if one works with a multi-stage set-up, and uses less electricity. A cross-current scrubber is suited to emissions with known maximum concentrations, thus allowing it to be dimensioned appropriately. In case of very high concentration peaks, for which the scrubber has not been dimensioned, the scrubbing liquid will be saturated before it reaches the bottom of the packing. This means that a part of the air will not be (fully) treated, with yield loss as a result.

Gas Scrubber with or without Built-in Device

Gas scrubbers can also be distinguished by the set-up of the wash section, e.g. with or without a built-in device. The built-in device could be a bulk or structured packing or a construction with plates or a rotating disk. The main layout can be further broken down as follows:

Gas scrubbers without built-in device:

- Spray towers: In spray towers the water is dispersed in fine droplets, normally via sprayers at the top of the scrubber, while the gas is fed from underneath – thus in counter-current. Set-up is also possible in co-current or cross-current formats. Can also be used as a dust scrubber.

- Jet scrubbers: In a jet scrubber, the gas and scrubbing liquid are brought into contact with one another in a co-current direction, in accordance with the workings of a water jet pump. In the wash section, the jet breaks down into droplets, which creates a large phase interface. In the next area, the gas and the liquid are separated.

- Venturi scrubber: A venturi scrubber consists of a converging section, a throat (the narrowest part of the venturi tube) and a diffuser. The gas flows through the venturi tube and reaches top speed in the throat section. Thereafter, the gas passes into the diffuser where the speed of the gas drops once again. The liquid is added to the gas flow either in the throat section or prior to it. Intensive mixing takes place between the gas and the liquid in the throat section of the venturi tube. Due to the high speed realised by the gas and liquid, the water is broken down into fine water droplets.Can also be used as a dust scrubber.

Gas scrubbers with built-in device:

- Plate column: A plate column is a column which is divided into segments by perforated plates. The perforations have been designed in a way that forces the to-be-cleaned gas to bubble through a sealed fluid layer on the plates, which is where absorption takes place.

- Packed columns: Scrubbers with packed columns are filled with structured or unstructured packing material. This material has a high specific surface area, which means a large phase interface is created between the gas and the liquid. The scrubbing liquid flows downwards in a thin film over the packing material, while the gas flows upwards through the remaining free space. In scrubbers with packed columns, the liquid and the gas do not disperse into one another.

- Rotation scrubber: In rotation scrubbers the scrubbing liquid is, via a fast-rotating spray, broken down into small droplets, whereby a large contact area is created between droplets and gas. As a result of the rotating sprayer, dust particles are forced to the sides of the scrubber and separated. Rotation scrubbers are primarily used as dust scrubbers.

- Ionisation scrubbers: These are a modified form of wet E filters. They are scrubbers with a built-in ionisation phase.

The compatibility of the various scrubber types is determined by the properties of the to-be-cleaned gas.

If it contains a lot of solid particles or other components that could lead to cake-forming and blockage, then a scrubber will be selected which is less sensitive to these factors - such a various scrubbers without built-in devices.

Another possibility is to install a multi-stage scrubbing system, where the various stages are designed to remove different components. Plate columns are primarily used in the chemicals industry. They are rarely used for environmental purposes due to the high investment costs.

Packed columns are normally used during absorption applications. One must choose between a bulk packing or a structured packing. Bulk packings are cheaper, have a lower specific surface area and a higher pressure drop. Structured packings, on the other hand, are a little bit more expensive than bulk packings, have a high specific surface area and a lower pressure drop. The choice between the two types of packing is determined by the to-be-treated gas stream. If there is a considerable risk of blockage due to dust and/or biological growth, then an open packing – which is easier to clean - will be used. In other cases, a packing with a smaller opening and a higher specific surface area will be used.

Layout according to type of scrubbing liquid

- Acid scrubber

- Alkaline scrubber

- Alkaline oxidizing scrubber

- Wet lime scrubber

- Bioscrubber

- Water scrubber: For substances that dissolve well in water, such as certain alcohols, no additional substances are added

- Oil scrubber: This is possible for lipophilic products such as, for example, halogenated solvents

Efficiency

Depending on the to-be-removed component, residual emission, scrubbing liquid and the type of application, yields in excess of 99% can be realised.

In asphalt plants, a yield of almost 98% has been registered for VOC's. In terms of odour reduction, the yield was maximum 23%. Considering the poor water solubility of VOC's in flue gases, it is expected that yields will deteriorate due to the low solubility, whereby the VOC will no longer be collected .

Boundary Conditions

- Flow rate: 50 – 500 000 Nm³/h

- Temperature: 5 - 80 °C

- Dust: < 10 mg/m³

Auxiliary Materials

- Water. Water use is determined by the in and out-going concentrations of gaseous components.

- Reagents: Acids, alkalis, bleach, peroxide etc. depending on used variant.

- Apart from water, no specific chemicals are needed for the removal of HC1 from flue gases.

Environmental Aspects

Waste water. In most cases, waste water needs to be purified. In certain cases it can be evaporated and reprocessed for the recuperation or recovery of products.

Acidic leachate will be partly drained (depending on pH). The leachate is supplemented by water. The released leachate must be treated prior to being discharged.

Energy use

Energy use lies between 0.2 – 1.0 kWh/1 000 Nm³/h (excluding ventilator) .

Cost Aspects

- *Investment*
  - 2 000 – 30 000 EUR for 1 000 Nm³/h (recirculation scrubber with pomp; costs greatly determined by application).
  - For the removal of HCl from flue gases from a chemicals company, with a flue gas flow rate up to 3 000 Nm³/h, the investment costs for a neutral washer (water) and a ventilator, amount to 62 500 EUR.

- *Operating costs*
  - Personnel costs: ca. 5 000 EUR per year (estimated at 4 hours per week).
  - Auxiliary and residual materials: Determined by in-going concentrations and required residual emissions.

- Ethanol case study
  - Flow rate: 13 000 Nm³/h
  - Single-stage counter-current scrubber in polyester
  - Washing liquid: water
  - Drainage: 0 -2 m³/h
  - Circulation pump: 3 kW
  - Investment costs: 85 000 EUR

Advantages and Disadvantages

- *Advantages*
  - Broad application spectrum;
  - Very high removal yields;
  - Compact installation and easy to maintain;
  - Relatively simple technology;
  - Can also be used to cool hot gas flows (quencher).

- *Disadvantages*
  - Waste water must be treated;

— Water and reagents used;

— When dust is simultaneously collected, drainage is necessary;

— Susceptible to frost;

— Depending on the location, a support construction may be necessary;

— Packing material could possibly be susceptible to blockage by dust (> 10 mg/m3) and fat;

— Pilot tests are often required for odour problems in order to evaluate attainability.

Applications

Broad range of applications in:

- Chemicals industry
- Waste incineration installations;
- Pharmaceutical industry
- Storage and transfer of chemicals
- Surface treatment

Synonyms, abbreviations and/or process names

Base scrubber

Removed Components

- HCl
- HF
- SO_2
- Cr_2O_7
- Cl_2
- Phenols
- Organic acids
- H_2S

Diagram

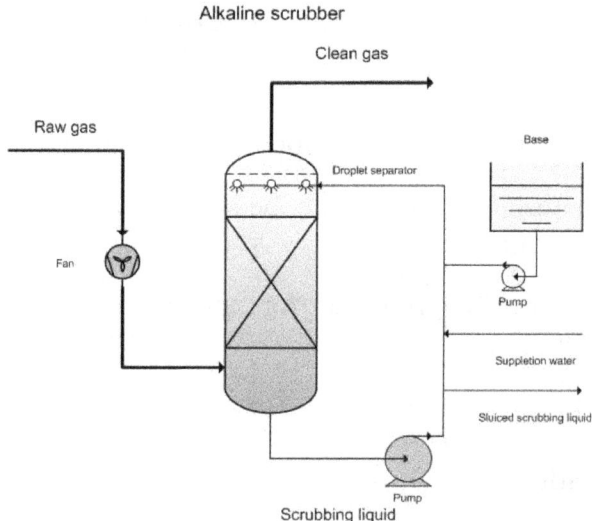

Process Description

For a general process description of a gas scrubber, please refer to 'gas scrubbing – general'. In an alkaline scrubber, acid-forming components are collected via neutralisation using a base as scrubbing liquid. This leads to the formation of salts, which can eventually be reprocessed. The drained water is purified and discharged into the sewer network.

Alkaline dosage occurs on the basis of pH control. The pH for an alkaline scrubber is normally kept between 8.5 and 9.5. The pH cannot be set very high, due to increased alkaline usage at higher pHs due to absorption of CO_2 in the water. From a pH above 10, the dissolved CO_2 will be present in the water as carbonate, whereby alkaline use will increase sharply. The calcium carbonate will also precipitate on the packing, whereby the pressure drop will increase. In order to avoid this, it is recommended that softened water be used for alkaline scrubbers.

Variants

For the different variants (counter-current, co-current or cross-current, with or without built-in device), please refer to 'gas scrubbing – general'.

Efficiency

- HCl: > 99 %; < 10 mg/Nm³

- HF: > 99 %; < 1 mg/Nm³

- SO_2:   > 99 %; < 40 mg/Nm³

- Phenols: > 90 %

- For biogas: H_2S concentrations of 1 000 – 10 000 ppm are reduced to 200 – 500 ppm which corresponds to a yield of 90 – 95 %.

Boundary Conditions

- Flow rate: 50 – 500 000 Nm³/h

- Temperature: 5 - 80 °C

- Dust: < 10 mg/m³

- HCl: 50 – 20 000 mg/Nm³

- HF: 50 – 1 000 mg/Nm³

- SO_2: 100 – 10 000 mg/Nm³

Auxiliary Materials

- Water

- Alkalis: Sodium hydroxide, sodium (bi-) carbonate etc.; pH regulation recommended.

Environmental Aspects

Waste water. In most cases, waste water needs to be purified. In certain cases it can be evaporated and reprocessed for recuperation or recovery of products.

Energy use

Energy use lies between 0.2 – 1.0 kWh/1 000 Nm³/h. Strongly determined by application .

Cost Aspects

- *Investment*

    - 2 000 – 30 000 EUR for 1 000 Nm³/h (strongly determined by application and implementation)

- *Operating costs*

    - Personnel costs: ca. 5 000 EUR per year (estimated at 4 hours per week)

    - Auxiliary and residual materials: Determined by in-going concentrations and required residual emissions

    - Cost aspects NaOH (29 %): ca. 210 EUR/ton, depending on volume

- Flue gas desulphurisation case study :

  — Household-waste incineration plant with a capacity of ca. 80 000 ton per year and a flue gas flow rate of 60 000 Nm³/h.

  — Investment cost of ca. 2 500 000 EUR.

  — Average concentration for SO_2 400 mg/Nm³.

  — Continuously operating installation: Emissions of 210 ton per year.

  — To reduce SO_2 emissions by 50%, the use of NaOH averages 48 kg/h or 1 236 kg/day. On an annual basis, this means ca. 7.5 ton per year per 1 000 Nm³/h or 1 575 EUR per year per 1 000 Nm³/h.

- Phenol-removal case study

  — Flow rate: 3 000 Nm³/h.

  — Single-stage counter-current scrubber in 316 stainless steel.

  — Temperature: Environment temperature.

  — Investment: 80 000 EUR.

  — Discharge: 30 l/h.

  — With later employed activated carbon filter in Ex set-up.

  — Circulation pump: 4 kW.

- Case study HC1 removal

  — Flow rate: 3 x 3 000 Nm³/h.

  — Single-stage counter-current scrubber in polyester.

  — Temperature: ambient temperature.

  — Investment: 62 500 EUR incl. ventilator.

  — Discharge: 300 l/h.

  — Circulation pump: 2 kW, ventilator: 1.5 kW.

- Biogas desulphurisation case study

  — Investment: between 80 000 and 150 000 EUR/1000 Nm³/hour and almost independent of treated volume

  — Sodium hydroxide-storage: 30 and 60 EUR/1000 Nm³/hour

  — Costs for biogas desulphurisation: Approx. 0.74 EUR per ton biomass

  — Operating costs: Primarily lye use, but also personnel, electricity and water.

Advantages and Disadvantages

- *Advantages*

  — Relatively compact;

  — Very high removal yields

  — Can be constructed in modules, multi-stage systems

- *Disadvantages*

  — Alkaline use, pH regulation recommended;

  — Waste water: the quantity can be restricted by checking the discharge for density and conductivity. If economically viable, in certain cases, the formed salts are reprocessed and re-used.

Applications

Is used for the removal of acid-forming components during incineration processes in:

  — Electricity plants

  — Waste combustion installations

  — Electro industry

  — Chemicals industry

  — Chlorine production

Adsorber

Many chemicals can be removed from exhaust gas also by using adsorber material. The flue gas is passed through a cartridge which is filled with one or several adsorber materials and has been adapted to the chemical properties of the components to be removed. This type of scrubber is sometimes also called dry scrubber. The adsorber material has to be replaced after its surface is saturated.

Note: adsorption is a surface phenomena, absorption involves the entire material. Ex: Activated carbon an adsorbent, used for the adsorption of odorous compounds.

Mercury Removal

Mercury is a highly toxic element commonly found in coal and municipal waste. Wet scrubbers are only effective for removal of soluble mercury species, such as oxidized mercury, Hg^{2+}. Mercury vapor in its elemental form, Hg^{0}, is insoluble in the scrubber slurry and not removed. Therefore, an additional process of Hg^{0} conversion is required to complete mercury capture. Usually halogens are added to the flue gas for this purpose. The type of coal burned as well as the presence of a selective catalytic reduction unit both affect the ratio of elemental to oxidized mercury in the flue gas and thus the degree to which the mercury is removed.

In July 2015, one study found that some mercury scrubbers installed on coal power plants inadvertently capture PAH (polycyclic aromatic hydrocarbons) emissions as well.

Scrubber Waste Products

One side effect of scrubbing is that the process only moves the unwanted substance from the exhaust gases into a liquid solution, solid paste or powder form. This must be disposed of safely, if it can not be reused.

For example, mercury removal results in a waste product that either needs further processing to extract the raw mercury, or must be buried in a special hazardous wastes landfill that prevents the mercury from seeping out into the environment.

As an example of reuse, limestone-based scrubbers in coal-fired power plants can produce a synthetic gypsum of sufficient quality that can be used to manufacture drywall and other industrial products.

Bacteria Spread

Poorly maintained scrubbers have the potential to spread disease-causing bacteria. The problem is a result of inadequate cleaning. For example, the cause of a 2005 outbreak of Legionnaires' disease in Norway was just a few infected scrubbers. The outbreak caused 10 deaths and more than 50 cases of infection.

Wet Scrubber

A wet scrubber or wet scrubber system is one type of scrubber that is used to remove harmful materials from industrial exhaust gases before they are released into the environment. It was the original type of scrubbing system, and utilizes a wet substance to remove acidic gases that contribute to acid rain.

When using a wet scrubber, gas is funneled through an area and sprayed with a wet substance. Water is used when dust and particulate matter is to be removed, but other chemicalscan be added. These chemicals are chosen to specifically react with certain airborne contaminants - generally acidic gases. This process adds significant amounts of vapour to the exhaust - which results in the exhaust looking like a white smoke when vented.

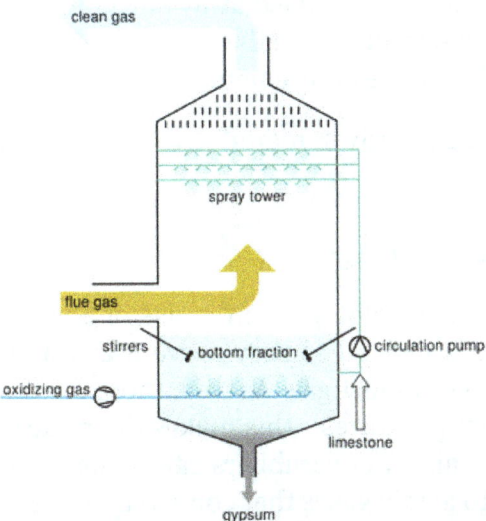

clean gas

spray tower

flue gas

stirrers   bottom fraction   circulation pump

oxidizing gas

limestone

gypsum

A diagram of a wet scrubber. This diagram shows a collection pool at the bottom
where the waste liquid is drained off, and the misters at the top of the chamber
where the liquid is misted onto the exhaust gasas it moves upwards.

One reason for the development of dry scrubbing was because the weight and volume
of this used spray was significant, which led to difficulties in storing and disposing of
the waste material.

Use

Wet scrubbers are a special device used to remove a variety of pollutants from exhaust
gas from furnaces or other devices. These devices use a scrubbing liquid to remove the
pollutants. The exhaust gas is moved through the scrubbing liquid - generally it is passed
through a chamber and the liquid is misted through the gas - and the gas emerges without
the contaminants and pollutants that existed before exposure to the scrubbing liquid.
When the gas is sprayed with the fluid, the heavier pollutants are pulled out of the gas and
attach to the liquid because of its chemical composition. As the gas is passed through the
cleaning mist, the contaminants are attracted to the mist and left behind.

Although misting is a common method of cleaning exhaust gas in wet scrubbing, a different design forces the gas to bubble through a pool of scrubbing fluid. The method for
removing the contaminants is mostly the same, however, as the contaminants bind to
the fluid as the gas is filtered through the pool. This leaves the gas clean as it comes out
and leaves the contaminants in the pool.

Regardless of which method of introducing the scrubbing fluid is used, most wet scrubbers are similar in design. A typical scrubber is composed of ductwork and a fan system
to force gas through its chambers. There is also a pump, and a collection area for used
scrubbing liquid and some method to bring the used fluid away from the cleaned gas.
The liquid sprayed through the exhaust collects at the bottom of the chamber where

the spraying occurs. This liquid is funneled away and collected for specialized disposal because of the potentially harmful materials contained in it. This liquid cannot simply be thrown away or reused because of its chemical content.

These scrubbers are used frequently in manufacturing plants that process propane and other types of natural gas.

Advantages and Disadvantages

There are advantages and disadvantages to the use of a wet scrubber. First and foremost, these scrubbers are beneficial as they prevent a wide range of pollutants from entering the air through the exhaust gas. As well, these units are fairly sturdy and can tolerate a wide range of temperatures - this makes them ideal for operation in almost any environment. Additionally, wet scrubbers can be used to remove a wide range of pollutants from mercury to acidic gases that contribute to acid rain.

Despite the advantages, there are a few drawbacks. These machines require frequent maintenance, and they can suffer from corrosion quite severely. If maintained and vented properly, these machines can be used for many years before they require replacement.

Design

The design of wet scrubbers or any air pollution control device depends on the industrial process conditions and the nature of the air pollutants involved. Inlet gas characteristics and dust properties (if particles are present) are of primary importance. Scrubbers can be designed to collect particulate matter and/or gaseous pollutants. The versatility of wet scrubbers allow them to be built in numerous configurations, all designed to provide good contact between the liquid and polluted gas stream.

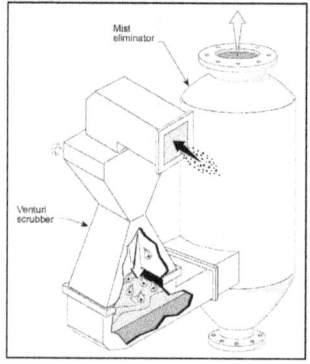

A venturi scrubber design. The mist eliminator for a venturi scrubber is often a separate device called a cyclonic separator

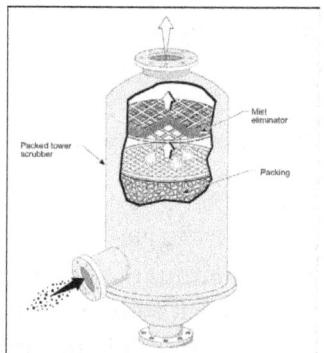

A packed bed tower design where the mist eliminator is built into the top of the structure. Various tower designs exist

Wet scrubbers remove dust particles by *capturing* them in liquid droplets. The droplets are then collected, the liquid *dissolving* or *absorbing* the pollutant gases. Any droplets that are in the scrubber inlet gas must be separated from the outlet gas stream by means of another device referred to as a mist eliminator or entrainmentseparator (these terms are interchangeable). Also, the resultant scrubbing liquid must be treated prior to any ultimate discharge or being reused in the plant.

A wet scrubber's ability to collect small particles is often directly proportional to the power input into the scrubber. Low energy devices such as spray towers are used to collect particles larger than 5 micrometers. To obtain high efficiency removal of 1 micrometer (or less) particles generally requires high-energy devices such as venturi scrubbers or augmented devices such as condensation scrubbers. Additionally, a properly designed and operated entrainment separator or mist eliminator is important to achieve high removal efficiencies. The greater the number of liquid droplets that are not captured by the mist eliminator, the higher the potential emission levels.

Wet scrubbers that remove gaseous pollutants are referred to as *absorbers*. Good gas-to-liquid contact is essential to obtain high removal efficiencies in absorbers. Various wet-scrubber designs are used to remove gaseous pollutants, with the packed tower and the plate tower being the most common.

If the gas stream contains both particulate matter and gases, wet scrubbers are generally the only single air pollution control device that can remove both pollutants. Wet scrubbers can achieve high removal efficiencies for either particles or gases and, in some instances, can achieve a high removal efficiency for both pollutants in the same system. However, in many cases, the best operating conditions for particles collection are the poorest for gas removal.

In general, obtaining high simultaneous gas and particulate removal efficiencies requires that one of them be easily collected (i.e., that the gases are very soluble in the liquid or that the particles are large and readily captured), or by the use of a scrubbing reagent such as lime or sodium hydroxide.

Components

Wet scrubber systems generally consist of the following components:

- Ductwork and fan system
- A saturation chamber (optional)
- Scrubbing vessel
- Entrainment separator or mist eliminator
- Pumping (and possible recycle system)
- Spent scrubbing liquid treatment and/or reuse system
- An exhaust stack

A typical wet scrubbing process can be described as follows:

- Hot flue gas from a furnace enters a saturator (if present) where gases are cooled and humidified prior to entering the scrubbing area. The saturator removes a small percentage of the particulate matter present in the flue gas.

- Next, the gas enters a venturi scrubber where approximately half of the gases are removed. Venturi scrubbers have a minimum particle removal efficiency of 95%.

- The gas flows through a second scrubber, a packed bed absorber, where the rest of the gases (and particulate matter) are collected.

- An entrainment separator or mist eliminator removes any liquid droplets that may have become entrained in the flue gas.

- A recirculation pump moves some of the spent scrubbing liquid back to the venturi scrubber where it is recycled and the remainder is sent to a treatment system.

- Treated scrubbing liquid is recycled back to the saturator and the packed bed absorber.

- Fans and ductwork move the flue gas stream through the system and eventually out the stack.

Categorization

Since wet scrubbers vary greatly in complexity and method of operation, devising categories into which all of them neatly fit is extremely difficult. Scrubbers for particle collection are usually categorized by the gas-side pressure drop of the system. Gas-side pressure drop refers to the pressure difference, or pressure drop, that occurs as the exhaust gas is pushed or pulled through the scrubber, disregarding the pressure that would be used for pumping or spraying the liquid into the scrubber.

Scrubbers may be classified *by pressure drop* as follows:

- *Low-energy scrubbers* have pressure drops of less than 12.7 cm (5 in) of water.

- *Medium-energy scrubbers* have pressure drops between 12.7 and 38.1 cm (5 and 15 in) of water.

- *High-energy scrubbers* have pressure drops greater than 37.1 cm (15 in) of water.

However, most scrubbers operate over a wide range of pressure drops, depending on their specific application, thereby making this type of categorization difficult.

Another way to classify wet scrubbers is by their *use* - to primarily collect either particulates or gaseous pollutants. Again, this distinction is not always clear since scrubbers can often be used to remove both types of pollutants.

Wet scrubbers can also be categorized by the manner in which the gas and liquid phases are brought into contact. Scrubbers are designed to use power, or energy, from the gas stream or the liquid stream, or some other method to bring the pollutant gas stream into contact with the liquid. These categories are given in Table.

Categories of wet collectors by energy source used for contact

Wet collector	Energy source used for gas-liquid contact
• Gas-phase contacting	• Gas stream
• Liquid-phase contacting	• Liquid stream
• Wet film	• Liquid and gas streams
• Combination	• Energy source:
○ Liquid phase and gas phase	○ Liquid and gas streams
○ Mechanically aided	○ Mechanically driven rotor

Material of Construction and Design

Corrosion can be a prime problem associated with chemical industry scrubbing systems. Fibre-reinforced plastic and dual keys are often used as most dependable materials of construction.

Venturi Scrubber

Particulate Scrubbers, often called Venturi Scrubbers, are effective at removing particulate from exhaust with high efficiency. Particulate scrubbers have several advantages compared to other dust collection equipment. They are able to handle gas streams that contain moisture and/or are high temperature, the overall size of the equipment is typically smaller, and they have the potential to remove pollutant gas at the same time as the particulate

Venturi scrubber systems incorporate a scrubbing vessel with a system fan, recycle pump, instrumentation and controls, mist eliminator, and exhaust stack. remove particles from gas streams by capturing the particles in liquid droplets and then separating the droplets from the gas stream. Particle laden exhaust gas enters the scrubber where it becomes in contact with a mist of tiny droplets of scrubbing fluid. The fluid is separated from the gas to remove the particulate. Collection efficiency is typically in excess of 99%.

Particulate Scrubber Applications

Pollution Systems' high quality design and construction also incorporate self cleaning features to maintain high on-stream performance. Some of the more common industrial applications of venturi scrubbers include Corn Processing Facilities, Food Manufacturing, Machining and Grinding exhaust, Fiberglass & Composite Industries, Utility Boilers, Foundries and Metal Finishing Operations.

Particle Collection

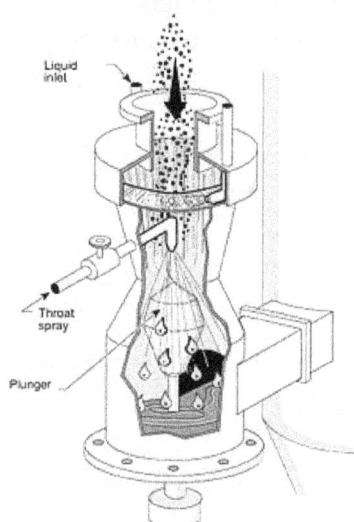

Adjustable-throat venturi scrubber with plunger

Venturis are the most commonly used scrubber for particle collection and are capable of achieving the highest particle collection efficiency of any wet scrubbing system. As the inlet stream enters the throat, its velocity increases greatly, atomizing and turbulently mixing with any liquid present.

The atomized liquid provides an enormous number of tiny droplets for the dust particles to impact on. These liquid droplets incorporating the particles must be removed from the scrubber outlet stream, generally by cyclonic separators.

Particle removal efficiency increases with increasing pressure drop because of increased turbulence due to high gas velocity in the throat. Venturis can be operated with pressure drops ranging from 12 to 250 cm (5 to 100 in) of water.

Most venturis normally operate with pressure drops in the range of 50 to 150 cm (20 to 60 in) of water. At these pressure drops, the gas velocity in the throat section is usually between 30 and 120 m/s (100 to 400 ft/s), or approximately 270 mph at the high end. These high pressure drops result in high operating costs.

The liquid-injection rate, or liquid-to-gas ratio (L/G), also affects particle collection. The proper amount of liquid must be injected to provide adequate liquid coverage over the throat area and make up for any evaporation losses. If there is insufficient liquid, then there will not be enough liquid targets to provide the required capture efficiency.

Most venturi systems operate with an L/G ratio of 0.4 to 1.3 l/m³ (3 to 10 gal/1000 ft³) (*Brady and Legatski 1977*). L/G ratios less than 0.4 l/m³ (3 gal/1000 ft³) are usually not sufficient to cover the throat, and adding more than 1.3 l/m³ (10 gal/1000 ft³) does not usually significantly improve particle collection efficiency.

Gas Collection

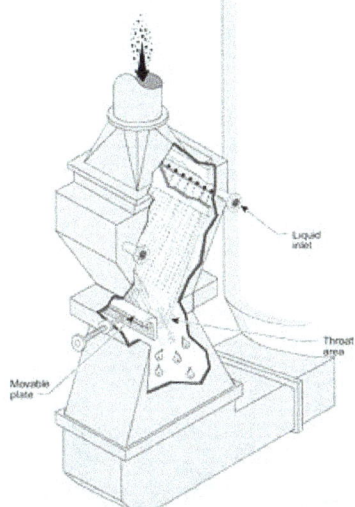

Liquid
inlet

Throat
area

Movable
plate

Adjustable-throat venturi scrubber with movable plate

Venturi scrubbers can be used for removing gaseous pollutants; however, they are not used when removal of gaseous pollutants is the only concern.

The high inlet gas velocities in a venturi scrubber result in a very short contact time between the liquid and gas phases. This short contact time limits gas absorption. However, because venturis have a relatively open design compared to other scrubbers, they are very useful for simultaneous gaseous and particulate pollutant removal, especially when:

- Scaling could be a problem

- A high concentration of dust is in the inlet stream

- The dust is sticky or has a tendency to plug openings

- The gaseous contaminant is very soluble or chemically reactive with the liquid

To maximize the absorption of gases, venturis are designed to operate at a different set of conditions from those used to collect particles. The gas velocities are lower and the liquid-to-gas ratios are higher for absorption.

For a given venturi design, if the gas velocity is decreased, then the pressure drop (resistance to flow) will also decrease and vice versa. Therefore, by reducing pressure drop, the gas velocity is decreased and the corresponding residence time is increased. Liquid-to-gas ratios for these gas absorption applications are approximately 2.7 to 5.3 l/m³ (20 to 40 gal/1000 ft³). The reduction in gas velocity allows for a longer contact time between phases and better absorption.

Increasing the liquid-to-gas ratio will increase the potential solubility of the pollutant in the liquid.

Though capable of some incidental control of volatile organic compounds (VOC), generally venturi scrubbers are limited to control PM (particulate matter) and high solubility gases.

Maintenance Problems

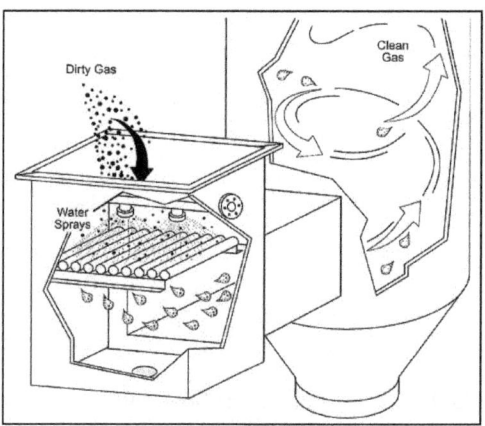

Venturi rod scrubber

The primary maintenance problem for venturi scrubbers is wear, or abrasion, of the scrubber shell because of high gas velocities. Gas velocities in the throat can reach speeds of 430 km/h (270 mph). Particles and liquid droplets traveling at these speeds can rapidly erode the scrubber shell.

Abrasion can be reduced by lining the throat with silicon carbide brick or fitting it with a replaceable liner. Abrasion can also occur downstream of the throat section. To reduce abrasion here, the elbow at the bottom of the scrubber (leading into the separator) can be flooded (i.e. filled with a pool of scrubbing liquid). Particles and droplets impact on the pool of liquid, reducing wear on the scrubber shell.

Another technique to help reduce abrasion is to use a precleaner (i.e., quench sprays or cyclone) to remove the larger particles.

The method of liquid injection at the venturi throat can also cause problems. Spray nozzles are used for liquid distribution because they are more efficient (have a more effective spray pattern) for liquid injection than weirs. However, spray nozzles can easily plug when liquid is recirculated. Automatic or manual reamers can be used to correct this problem. However, when heavy liquid slurries (either viscous or particle-loaded) are recirculated, open-wear injection is often necessary.

Cyclonic Spray Scrubber

Cyclonic spray scrubbers are an air pollution control technology. They use the features of both the dry cyclone and the spray chamber to remove pollutants from gas streams.

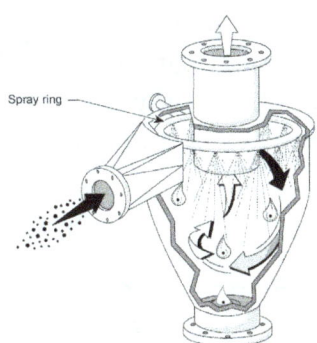

Irrigated cyclone scrubber

Generally, the inlet gas enters the chamber tangentially, swirls through the chamber in a corkscrew motion, and exits. At the same time, liquid is sprayed inside the chamber. As the gas swirls around the chamber, pollutants are removed when they impact on liquid droplets, are thrown to the walls, and washed back down and out.

Cyclonic scrubbers are generally low- to medium-energy devices, with pressure drops of 4 to 25 cm (1.5 to 10 in) of water. Commercially available designs include the irrigated cyclone scrubber and the cyclonic spray scrubber.

In the irrigated cyclone (Figure 1), the inlet gas enters near the top of the scrubber into the water sprays. The gas is forced to swirl downward, then change directions, and return upward in a tighter spiral. The liquid droplets produced capture the pollutants, are eventually thrown to the side walls, and carried out of the collector. The "cleaned" gas leaves through the top of the chamber.

The cyclonic spray scrubber (Figure 2) forces the inlet gas up through the chamber from a bottom tangential entry. Liquid sprayed from nozzles on a center post (manifold) is directed toward the chamber walls and through the swirling gas. As in the irrigated cyclone, liquid captures the pollutant, is forced to the walls, and washes out. The "cleaned" gas continues upward, exiting through the straightening vanes at the top of the chamber.

This type of technology is a part of the group of air pollution controls collectively referred to as wet scrubbers.

Particulate Collection

Cyclonic spray scrubbers are more efficient than spray towers, but not as efficient as venturi scrubbers, in removing particulate from the inlet gas stream. Particulates larger than 5 µm are generally collected by impaction with 90% efficiency. In a simple spray tower, the velocity of the particulates in the gas stream is low: 0.6 to 1.5 m/s (2 to 5 ft/s).

By introducing the inlet gas tangentially into the spray chamber, the cyclonic scrubber increases gas velocities (thus, particulate velocities) to approximately 60 to 180 m/s

(200 to 600 ft/s). The velocity of the liquid spray is approximately the same in both devices. This higher particulate-to-liquid relative velocity increases particulate collection efficiency for this device over that of the spray chamber. Gas velocities of 60 to 180 m/s are equivalent to those encountered in a venturi scrubber.

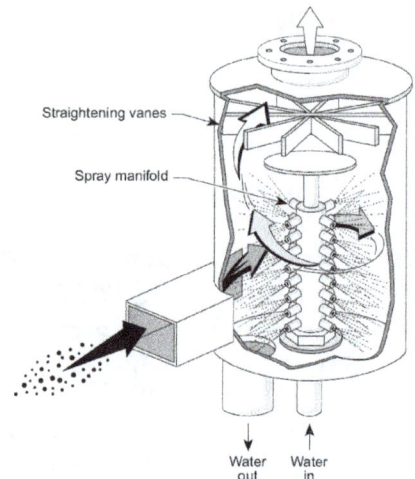

Cyclonic spray scrubber

However, cyclonic spray scrubbers are not as efficient as venturi scrubbers because they are not capable of producing the same degree of useful turbulence.

Gas Collection

High gas velocities through these devices reduce the gas-liquid contact time, thus reducing absorption efficiency. Cyclonic spray scrubbers are capable of effectively removing some gases; however, they are rarely chosen when gaseous pollutant removal is the only concern.

Maintenance Problems

The main maintenance problems with cyclonic scrubbers are nozzle plugging and corrosion or erosion of the side walls of the cyclone body. Nozzles have a tendency to plug from particulates that are in the recycled liquid and/or particulates that are in the gas stream. The best solution is to install the nozzles so that they are easily accessible for cleaning or removal.

Due to high gas velocities, erosion of the side walls of the cyclone can also be a problem. Abrasion-resistant materials may be used to protect the cyclone body, especially at the inlet.

Mechanically Aided Scrubber

Mechanically aided scrubbers are a form of pollution control technology. This type of technology is a part of the group of air pollution controls collectively referred to as wet scrubbers.

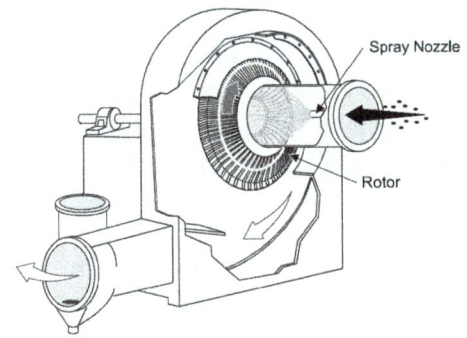

Centrifugal fan scrubber

In addition to using liquid sprays or the exhaust stream, scrubbing systems can use motors to supply energy. The motor drives a rotor or paddles which, in turn, generate water droplets for gas and particle collection.

Systems designed in this manner have the advantage of requiring less space than other scrubbers, but their overall power requirements tend to be higher than other scrubbers of equivalent efficiency. Significant power losses occur in driving the rotor. Therefore, not all the power used is expended for gas–liquid contact.

Types

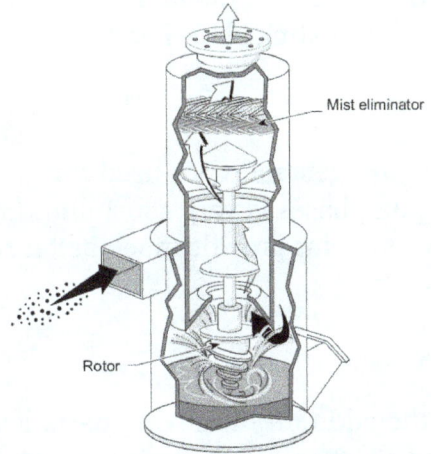

Induced spray scrubber

There are fewer mechanically aided scrubber designs available than liquid- and gas-phase contacting collector designs. Two are more common: *centrifugal fan scrubbers* and *mechanically induced spray scrubbers*.

A *centrifugal-fan scrubber* can serve as both an air mover and a collection device.

Figure shows such a system, where water is sprayed onto the fan blades cocurrently with the moving exhaust gas. Some gaseous pollutants and particles are initially removed as they pass over the liquid sprays.

The liquid droplets then impact on the blades to create smaller droplets for additional collection targets. Collection can also take place on the liquid film that forms on the fan blades. The rotating blades force the liquid and collected particles off the blades. The liquid droplets separate from the gas stream because of their centrifugal motion.

Centrifugal-fan collectors are the most compact of the wet scrubbers since the fan and collector comprise a combined unit. No internal pressure loss occurs across the scrubber, but a power loss equivalent to a pressure drop of 10.2 to 15.2 cm (4 to 6 in) of water occurs because the blower efficiency is low.

Another mechanically aided scrubber, the *induced-spray*, consists of a whirling rotor submerged in a pool of liquid. The whirling rotor produces a fine droplet spray. By moving the process gas through the spray, particles and gaseous pollutants can subsequently be collected.

Particle Collection

Mechanically aided scrubbers are capable of high collection efficiencies for particles with diameters of 1 μm or greater. However, achieving these high efficiencies usually requires a greater energy input than those of other scrubbers operating at similar efficiencies. In mechanically aided scrubbers, the majority of particle collection occurs in the liquid droplets formed by the rotating blades or rotor.

Gas Collection

Mechanically aided scrubbers are generally not used for gas absorption. The contact time between the gas and liquid phases is very short, limiting absorption. For gas removal, several other scrubbing systems provide much better removal per unit of energy consumed.

Maintenance Problems

As with almost any device, the addition of moving parts leads to an increase in potential maintenance problems. Mechanically aided scrubbers have higher maintenance costs than other wet collector systems. The moving parts are particularly susceptible to corrosion and fouling. In addition, rotating parts are subject to vibration-induced fatigue or wear, causing them to become unbalanced. Corrosion-resistant materials for these scrubbers are very expensive; therefore, these devices are not used in applications where corrosion or sticky materials could cause problems.

Particle Collection in Wet Scrubbers

Wet scrubbers capture relatively small dust particles with large liquid droplets. In most wet scrubbing systems, droplets produced are generally larger than 50 micrometres (in the 150 to 500 micrometres range). As a point of reference, human hair ranges in diameter from 50 to 100 micrometres. The size distribution of particles to be collected is source specific.

For example, particles produced by mechanical means (crushing or grinding) tend to be large (above 10 micrometres); whereas, particles produced from combustion or a chemical reaction will have a substantial portion of small (less than 5 micrometres) and submicrometre particles.

The most critical sized particles are those in the 0.1 to 0.5 micrometres range because they are the most difficult for wet scrubbers to collect.

Droplets are produced by several methods:

1. Injecting liquid at high pressure through specially designed nozzles;

2. Aspirating the particle-laden gas stream through a liquid pool;

3. Submerging a whirling rotor in a liquid pool.

These droplets collect particles by using one or more of several collection mechanisms such as impaction, direct interception, diffusion, electrostatic attraction, condensation, centrifugal force and gravity. However, impaction and diffusion are the main ones.

Impaction

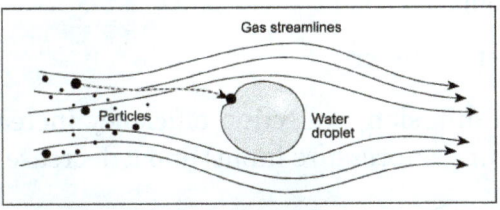

Impaction

In a wet scrubbing system, dust particles will tend to follow the streamlines of the exhaust stream. However, when liquid droplets are introduced into the exhaust stream, particles cannot always follow these streamlines as they diverge around the droplet. The particle's mass causes it to break away from the streamlines and impact or hit the droplet.

Impaction increases as the diameter of the particle increases and as the relative velocity between the particle and droplets increases. As particles get larger they are less likely to follow the gas streamlines around droplets. Also, as particles move faster relative to the liquid droplet, there is a greater chance that the particle will hit a droplet. Impaction is the predominant collection mechanism for scrubbers having gas stream velocities greater than 0.3 m/s (1 ft/s) (*Perry 1973*).

Most scrubbers operate with gas stream velocities well above 0.3 m/s. Therefore, at these velocities, particles having diameters greater than 1.0 µm are collected by this mechanism. Impaction also increases as the size of the liquid droplet decreases because the presence of more droplets within the vessel increases the probability that particles will impact on the droplets.

Diffusion

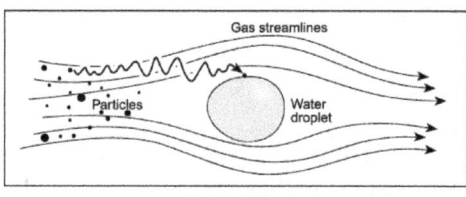

Diffusion

Very small particles (less than 0.1 µm in diameter) experience random movement in an exhaust stream. These particles are so tiny that they are bumped by gas molecules as they move in the exhaust stream. This bumping, or bombardment, causes them to first move one way and then another in a random manner, or to diffuse, through the gas. This irregular motion can cause the particles to collide with a droplet and be collected. Because of this, diffusion is the primary collection mechanism in wet scrubbers for particles smaller than 0.1 µm.

The rate of diffusion depends on the following:

1. The relative velocity between the particle and droplet;

2. The particle diameter;

3. The liquid-droplet diameter.

For both impaction and diffusion, collection efficiency increases with an increase in relative velocity (liquid- or gas-pressure input) and a decrease in liquid-droplet size.

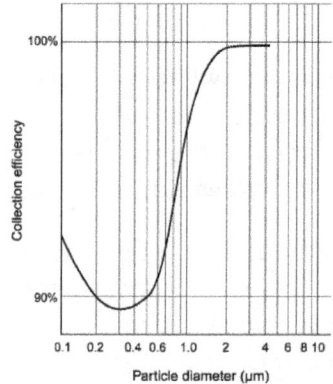

ypothetical curve illustrating relationship between particle size
and collection efficiency for a typical wet scrubber

However, collection by diffusion increases as particle size decreases. This mechanism enables certain scrubbers to effectively remove the very tiny particles (less than 0.1 μm).

In the particle size range of approximately 0.1 to 1.0 μm, neither of these two collection mechanisms (impaction or diffusion) dominates.

Other Collection Mechanisms

In recent years, some scrubber manufacturers have utilized other collection mechanisms such as electrostatic attraction and condensation to enhance particle collection without increasing power consumption.

In electrostatic attraction, particles are captured by first inducing a charge on them. Then, the charged particles are either attracted to each other, forming larger, easier-to-collect particles, or they are collected on a surface.

Condensation of water vapor on particles promotes collection by adding mass to the particles. Other mechanisms such as gravity, centrifugal force, and direct interception slightly affect particle collection.

Baffle Spray Scrubber

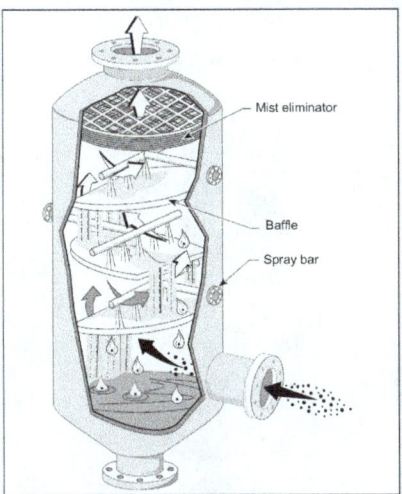

Baffle spray scrubber

Baffle spray scrubbers are a technology for air pollution control. They are very similar to spray towers in design and operation. However, in addition to using the energy provided by the spray nozzles, baffles are added to allow the gas stream to atomize some liquid as it passes over them.

A simple baffle scrubber system is shown in Figure. Liquid sprays capture pollutants and also remove collected particles from the baffles. Adding baffles slightly increases the pressure drop of the system.

This type of technology is a part of the group of air pollution controls collectively referred to as wet scrubbers.

A number of wet-scrubber designs use energy from both the gas stream and liquid stream to collect pollutants. Many of these combination devices are available commercially.

A seemingly unending number of scrubber designs have been developed by changing system geometry and incorporating vanes, nozzles, and baffles.

Particle collection

These devices are used much the same as spray towers - to preclean or remove particles larger than 10 μm in diameter. However, they will tend to plug or corrode if particle concentration of the exhaust gas stream is high.

Gas collection

Even though these devices are not specifically used for gas collection, they are capable of a small amount of gas absorption because of their large wetted surface.

Ejector Venturi Scrubber

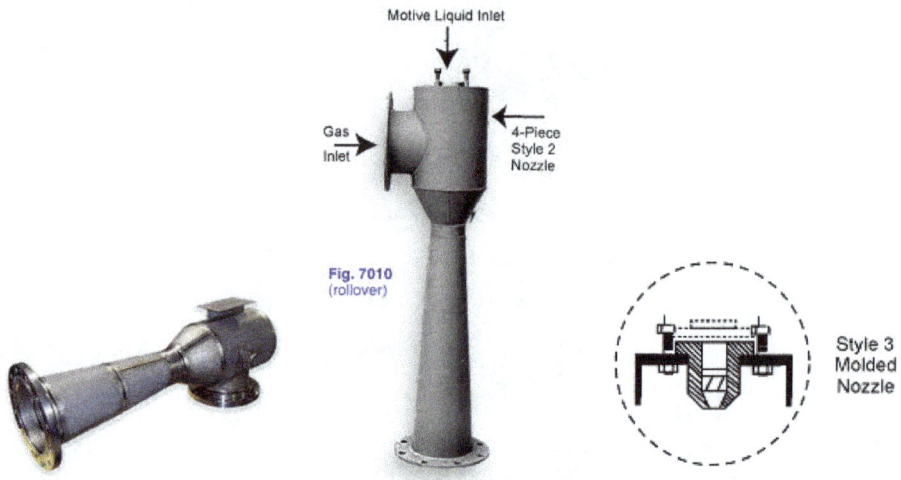

Ejector Venturi Gas Scrubbers are very effective at removing noxious gases, particulates, odors, fumes, and dusts from gas streams. Particulate contaminants are removed through impaction by the high-velocity spray of scrubbing liquid. Gases and odors are eliminated through absorption and/or chemical reaction between the gases and scrubbing liquid. When properly matched to the application, these scrubbers, by their nature, are better able to cope with the high temperatures, heavy contaminant loads, and corrosive conditions often encountered.

Pumping Characteristics

The pumping characteristics, draft vs capacity, of any given size unit can be varied with any given nozzle size by changing the pressure and correspondingly the quantity of water pumped. They also can be varied by keeping the liquid rate constant and changing the pressure. In the latter, changes in nozzle orifice size are required, and the available draft increases as the nozzle pressure increases, as illustrated in Figure. With the characteristics arranged in the manner shown in Figure, the pump horsepower required is directly proportional to the nozzle pressure. As a result, these characteristics illustrate the effects that the higher draft requirements for pumping the gas to or from the scrubber have upon nozzle pressure and pump horsepower. In a manner similar to other types of jet pumps, the gas pumping efficiency of the scrubber improves as the unit gets larger, especially in the range from 3 up to 12 in. The nozzle size and pressure combination used for any given application is de pendent upon the capacity, draft, liquid to gas ratio, and water drop size characteristics required for attaining the de sired scrubbing performance.

Drop Size Determination

The scrubbing action accomplished in these units is related to the size and quantity of the water drops produced by the spiral type nozzles. The size and number of drops for the various scrubbers from 3 up to 42 in. diam were visually observed and determined by photographing numerous segments of the spray from each of the nozzles with the single six micro-second light flash from a General Radio Strobotac. The photographic negatives were then enlarged so that the drops were 20 times normal size, as illustrated in Figure. This was done by means of a Lietz Vertical Projection Microscope. 1 This degree of magnification permitted sizing and counting the drops in the various segments of the spray. By means of statistical methods, the median and mean drop size characteristics by number and by weight for the entire spray from each size nozzle were determined. The mean drop size by number for all of the nozzles used in the 3 to 42 in.-scrubbers were found to be in the range from 800 microns for the smaller nozzles at low pressures to 300 microns for the very large nozzles at higher pressures. The size distribution characteristics of the drops produced by nozzles used in 4 and 12 in. scrubbers and how they are influenced by nozzle pressure is illustrated in Figures. In general, the higher pressure produces the smaller drop sizes, which is in general agreement with previous work reported by others on nozzles.6 On the other hand, it was observed that the larger nozzles produced the smaller drop sizes, which is not in general agreement with the work re ported by others on pressure nozzles.

Theoretical Equations

The theoretical collection efficiency as applied to wet type scrubbers has been derived, discussed, and reported by others in numerous papers. The simplified equation applicable to a hypothetical ejector venturi unit having a uniform drop size and a gas having a uniform dust particle size is

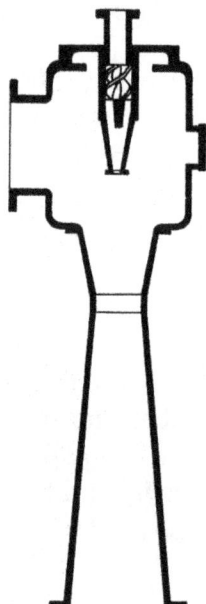

Cross-section of typical S&K ejector Venturi Scrubber

$$E = 1 - e^{\left(-30.5\, \eta_t\, PL/D_b\right)}$$

Where

n_t = target efficiency, %

P = effective scrubbing length, feet

L = liquid to gas ratio, gpm/1000 cu ft

D_b = water drop size, microns

The target or impaction efficiency of a single water drop upon a single dust particle has also been derived, discussed, and reported by others. This efficiency is of primary importance in estimating or predicting the over-all collection efficiency of the wet type collector. Generally, if preliminary calculations yield very low target efficiencies below 0.30, it is very probable that the collection efficiency and scrubbing action will be un-satisfactory. Various target efficiency parameters have been studied theoretically by various investigators, and the one in more common usage for the wet collectors is the equation presented by Langmuir and Blodgett.

$$n_t = f\left(\left(D_p 2_p U/18_u D_b\right)\right)$$

D_p = dust particle

p_p = particle density

U = relative velocity between water drops and the gas

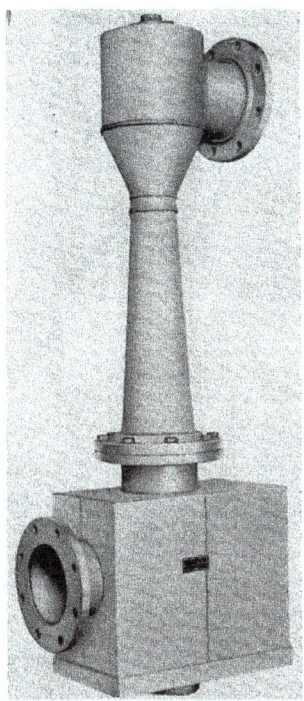

Ejector venturi scru bber with inertia im pact separator

u   =   viscosity of gas

D_b =   water drop size

Theoretical Efficiencies

The actual collection efficiency of an ejector venturi scrubber cannot be determined as simply as presented above. The analytical determination of the actual scrubbing efficiency is extremely complicated and probably almost impossible from a practical viewpoint due to the following conditions:

(1) The majority of the drops are far from being uniform in size and are in the range from 200 to 1700 microns as illustrated in Figures.

(2) The majority of the inlet particle sizes for any given dust probably varies over a size range of at least 10 to one.

(3) The actual path length a water drop travels in scrubbing a gas is probably impossible to determine with any reasonable degree of accuracy.

(4) The target or impaction relation ships for shapes other than simple geometric forms are probably unknown.

(5) The relative velocity between the water drops and the dust particles is not constant throughout the length of the scrubber and also extremely difficult to determine.

In spite of the complexities involved, the simplified theoretical collection efficiency and target efficiency equations presented above can be used to determine if a significant degree of scrubbing on a weight basis is attainable. By applying the previously mentioned target efficiency equation to a 12 in. scrubber handling a dust in the range of one to 10 microns with a density of 5.12 gm/cc values in range of 40 to 98% were obtained with nozzle pressures from 20 to 120 psi, as illustrated on Figure. The values shown were determined by using a mean drop size of 500 microns at the higher pressure and 800 microns at the lower pressure. The target efficiency values obtained were based upon the relationship and data for spherical shapes developed by Langmuir and Blodgett. The relative velocity used in the equation is a rough approximation and was obtained by the difference between the velocity of the water leaving the nozzle and the average velocity of the gas passing through the divergent portion of the venturi. The relative velocity values used are in the range of 24 fps at the lower nozzle pressure and 48 fps at the higher pressure. Depending upon the numerous controllable factors that effect the pumping capacity and efficiency, the inlet and discharge duct velocities for the many different scrubber applications and conditions can vary from a low value of 140 fpm at high draft to a high value of 3300 fpm with very low draft. As a result, when using the method described for obtaining the relative velocities, one may encounter many instances where the relative velocities are higher than mentioned and a few vice versa. Although the exact value and method of calculation of the relative velocities used are subject to question, they are for the majority of applications felt to be conservative, and therefore do serve their purpose of permitting the determination of the general picture of the theoretical efficiencies that can be obtained in an ejector venturi scrubber. The target efficiency values shown on Figure indicate that satisfactory scrubbing efficiencies are theoretically possible for particle sizes down to one micron diam. The mean drop sizes at the lower pressure and at the higher pressure are in the range of 800 and 500 microns respectively for the majority of the various size scrubber nozzles.

Consequently, the target efficiencies shown are also approximately that which would be obtained in other size units from 4 up to 42 in. for dust having a particle density of approximately 5 gm/cc and a generally spherical or cubic shape. Observation of the target efficiency equation indicates that higher efficiencies will be obtained for dusts having higher particle densities. Langmuir and Blodgett's data show that the relationship between target efficiency and the parameter in parenthesis, in the aforementioned equation, is far from linear. Consequently, changes in particle density from 5.12 gm/cc to a value of 1 or 2 gm/cc do not produce proportional or major changes in the target efficiency. Due to the relationship between target efficiency, collection efficiency, particle density, and particle size very little change in the theoretical collection efficiency occurs for changes in the particle density for orders of magnitude up to four times and for particle sizes of eight microns and above. Conversely, for particles in the range of one micron size the theoretical equation and data indicate that magnitude changes in particle density of two can cause major changes in the target and theoretical collection efficiencies.

The corresponding theoretical collection efficiencies for a 12 in. unit that are obtained using the illustrated target efficiencies, are shown on Figure. These values were obtained by assuming that the effective scrubbing length is equal to the length of the venturi throat and the divergent discharge taper. A liquid to gas ratio of 40 gpm/1000 cu ft was used. This value is very close to that which we have experimental data on. The data on the Figure show that from a theoretical view point, very significant and satisfactory collection efficiencies can be obtained in a 12 in. unit with the L/G ratio of 40 gpm/1000 cu ft at particle sizes of four microns and above. The effective scrubbing length varies considerably from one size to the next. As a result, the smaller units may have poorer collection efficiencies especially at the particle sizes less than four microns. On the other hand, larger sizes will have better collection efficiencies. From the collection efficiency equation, it can be seen that theoretically the same efficiency can be obtained in the smaller units as in the larger units by increasing the liquid to gas ratio. On the other hand, the same efficiency shown for the 12 in. unit can theoretically be obtained with lower liquid to gas rates on the larger units.

Carbon Dioxide Scrubber

A CO_2 Scrubber removes carbon dioxide from the air. Although large scale CO_2 scrubbers remain largely in the theoretical and planning stages of development, many environmentalists and scientists question the usefulness of this technology.

Working of CO_2 Scrubber

While there are many prototypes of CO_2 scrubbers, most use similar technology to remove carbon dioxide from the air. Although the technology to create the CO_2 scrubber is advanced, the function of the machine is easy to follow.

1. Air polluted with carbon dioxide is pumped into the CO_2 scrubber.

2. The air comes into contact with an ion exchange resin, which attracts the carbon dioxide molecules.

3. The cleaned air is then pumped out of the CO_2 scrubber.

The ion exchange resin must be cleaned periodically to retain its usefulness. Currently, humid air is used to remove the carbon from the scrubber. The carbon must then be discarded or reused. One possible use is to pump the collected carbon dioxide into a greenhouse to encourage quicker plant growth. Other scientists see potential in growing algae with the collected carbon, which could then be used as food or fertilizer.

Advantages of the CO_2 Scrubber

There are several advantages of the CO_2 scrubber that makes many people interested in further developing the technology.

- A CO_2 scrubber can help protect the economy by diminishing the need to reduce the use of fossil fuels, which remain less expensive than renewable and alternative energy sources.

- Finding ways to reduce carbon in the atmosphere, some argue, is simply pragmatic, because most consumers and corporations are not willing to reject fossil fuel use.

- Using a CO_2 scrubber will reduce the amount of greenhouse gases that lead to global warming.

- It is estimated that the amount of carbon in the atmosphere is nearly 40 percent higher than before the Industrial Revolution, so some environmentalists argue that scrubber technologies are needed regardless of advances in alternative energy sources.

Environmentalist Opposition to the Scrubber

While the CO_2 scrubber help resolve some short term environmental problems, many environmentalists are concerned that embracing this technology will just allow consumers to disregard the need to develop alternative energy sources. They argue that consumers and government agencies should be seeking ways to permanently diminish the use of fossil fuel and carbon emissions, not reduce the harmful side affects of this consumerism. In general, these groups argue people should be interested in ways to save energy, not ways to reduce the impact of rampant consumerism.

Other technologies such as carbon sequestration and carbon capture have also been rejected by many environmental groups for the same reason. These groups argue that controlling carbon emissions still does not address other problems inherent in fossil fuel use. These problems, such as a dwindling supply and ecosystem pollution, would not be resolved through CO_2 scrubber or other carbon technologies. Instead, they argue, scientists should be using solar power to decrease global warming and other renewable technologies.

The CO_2 scrubber is also currently too expensive for widespread use. While scientists argue that the benefits of reducing carbon dioxide in the air more than compensates for the technology's currently high price, it is unclear who would purchase the machines. Some think that money gained from carbon credits could be used.

There is also some question as to the environmental impact of the CO_2 scrubber. Collecting one ton of carbon dioxide is only equivalent to a single plane flight from London to New York. This means that millions of units may have to be made in order for a significant change to occur. There is also some debate as to how to dispose of the carbon that is collected in the scrubber.

Technologies

Amine Scrubbing

The dominant application for CO_2 scrubbing is for removal of CO_2 from the exhaust of coal- and gas-fired power plants. Virtually the only technology being seriously evaluated involves the use of various amines, e.g. monoethanolamine. Cold solutions of these organic compounds bind CO_2, but the binding is reversed at higher temperatures:

$$CO_2 + 2\ HOCH_2CH_2NH_2 \leftrightarrow HOCH_2CH_2NH_3^+ + HOCH_2CH_2NHCO_2^-$$

As of 2009, this technology has only been lightly implemented because of capital costs of installing the facility and the operating costs of utilizing it.

Minerals and Zeolites

Several minerals and mineral-like materials reversibly bind CO_2. Most often, these minerals are oxides, and often the CO_2 is bound as carbonate. Carbon dioxide reacts with quicklime (calcium oxide) to form limestone (calcium carbonate), in a process called carbonate looping. Other minerals include serpentinite, a magnesium silicate hydroxide, and olivine. Molecular sieves also function in this capacity.

Various scrubbing processes have been proposed to remove CO_2 from the air, or from flue gases. These usually involve using a variant of the Kraft process. Scrubbing processes may be based on sodium hydroxide. The CO_2 is absorbed into solution, transferred to lime via a process called causticization and released in a kiln. With some modifications to the existing processes, mainly an oxygen-fired kiln, the end result is a concentrated stream of CO_2 ready for storage or use in fuels. An alternative to this thermo-chemical process is an electrical one in which a nominal voltage is applied across the carbonate solution to release the CO_2. While simpler, this electrical process consumes more energy as it splits water at the same time. Since it depends on electricity, the electricity needs to be renewable, like PV. Otherwise the CO_2 produced during electricity production has to be taken into account. Early incarnations of air capture used electricity as the energy source; hence, were dependent on a carbon-free source. Thermal air capture systems use heat generated on-site, which reduces the inefficiencies associated with off-site electricity production, but of course it still needs a source of (carbon-free) heat. Concentrated solar power is an example of such a source.

Sodium Hydroxide

Zeman and Lackner outlined a specific method of air capture.

First, CO_2 is absorbed by an alkaline NaOH solution to produce dissolved sodium carbonate. The absorption reaction is a gas liquid reaction, strongly exothermic, here:

$$2NaOH(aq) + CO_2(g) \rightarrow Na_2CO_3(aq) + H_2O(l)$$

$$Na_2CO_3(aq) + Ca(OH)_2(s) \rightarrow 2NaOH(aq) + CaCO_3(s)$$

$$\Delta H^\circ = -5.3 \text{ kJ/mol}$$

Causticization is performed ubiquitously in the pulp and paper industry and readily transfers 94% of the carbonate ions from the sodium to the calcium cation. Subsequently, the calcium carbonate precipitate is filtered from solution and thermally decomposed to produce gaseous CO_2. The calcination reaction is the only endothermic reaction in the process and is shown here:

$$CaCO_3(s) \rightarrow CaO(s) + CO_2(g)$$

$$\Delta H^\circ = +179.2 \text{ kJ/mol}$$

The thermal decomposition of calcite is performed in a lime kiln fired with oxygen in order to avoid an additional gas separation step. Hydration of the lime (CaO) completes the cycle. Lime hydration is an exothermic reaction that can be performed with water or steam. Using water, it is a liquid/solid reaction as shown here:

$$CaO(s) + H_2O(l) \rightarrow Ca(OH)_2(s)$$

$$\Delta H^\circ = -64.5 \text{ kJ/mol}$$

Lithium Hydroxide

Other strong bases such as soda lime, sodium hydroxide, potassium hydroxide, and lithium hydroxide are able to remove carbon dioxide by chemically reacting with it. In particular, lithium hydroxide was used aboard spacecraft, such as in the Apollo program, to remove carbon dioxide from the atmosphere. It reacts with carbon dioxide to make lithium carbonate. Recently lithium hydroxide absorbent technology has been adapted for use in anesthesia machines. Anesthesia machines which provide life support and inhaled agents during surgery typically employ a closed circuit necessitating the removal of carbon dioxide exhaled by the patient. Lithium hydroxide may offer some safety and convenience benefits over the older calcium based products.

$$2 \text{ LiOH(s)} + 2 \text{ H}_2\text{O(g)} \rightarrow 2 \text{ LiOH}\cdot\text{H}_2\text{O(s)}$$

$$2 \text{ LiOH}\cdot\text{H}_2\text{O(s)} + \text{CO}_2\text{(g)} \rightarrow \text{Li}_2\text{CO}_3\text{(s)} + 3 \text{ H}_2\text{O(g)}$$

The net reaction being:

$$2\text{LiOH(s)} + \text{CO}_2\text{(g)} \rightarrow \text{Li}_2\text{CO}_3\text{(s)} + \text{H}_2\text{O(g)}$$

Lithium peroxide can also be used as it absorbs more CO_2 per unit weight with the added advantage of releasing oxygen.

Regenerative Carbon Dioxide Removal System

The regenerative carbon dioxide removal system (RCRS) on the space shuttle orbiter used a two-bed system that provided continuous removal of carbon dioxide without expendable products. Regenerable systems allowed a shuttle mission a longer stay in space without having to replenish its sorbent canisters. Older lithium hydroxide (LiOH)-based systems, which are non-regenerable, were replaced by regenerable metal-oxide-based systems. A system based on metal oxide primarily consisted of a metal oxide sorbent canister and a regenerator assembly. It worked by removing carbon dioxide using a sorbent material and then regenerating the sorbent material. The metal-oxide sorbent canister was regenerated by pumping air at approximately 400 °F (204 °C) through it at a standard flow rate of 7.5 cu ft/min (0.0035 m³/s) for 10 hours.

Activated Carbon

Activated carbon can be used as a carbon dioxide scrubber. Air with high carbon dioxide content, such as air from fruit storage locations, can be blown through beds of activated carbon and the carbon dioxide will adsorb onto the activated carbon. Once the bed is saturated it must then be "regenerated" by blowing low carbon dioxide air, such as ambient air, through the bed. This will release the carbon dioxide from the bed, and it can then be used to scrub again, leaving the net amount of carbon dioxide in the air the same as when the process was started.

Metal-organic Frameworks (MOFs)

Metal-organic frameworks are one of the most promising new technologies for carbon dioxide capture and sequestration via adsorption. Although no large-scale commercial technology exists nowadays, several research studies have indicated the great potential that MOFs have as a CO_2 adsorbent. Its characteristics, such as pore structure and surface functions can be easily tuned to improve CO_2 selectivity over other gases.

A MOF could be specifically designed to act like a CO_2 removal agent in post-combustion power plants. In this scenario, the flue gas would pass through a bed packed with a MOF material, where CO_2 would be stripped. After saturation is reached, CO_2 could be desorbed by doing a pressure or temperature swing. Carbon dioxide could then be compressed to supercritical conditions in order to be stored underground or utilized in enhanced oil recovery processes. However, this is not possible in large scale yet due to several difficulties, one of those being the production of MOFs in great quantities.

Another problem is the availability of metals necessary to synthesize MOFs. In a hypothetical scenario where these materials are used to capture all CO_2 needed to avoid global warming issues, such as maintaining a global temperature rise less than 2°C above the pre-industrial average temperature, we would need more metals than are available on Earth. For example, to synthesize all MOFs that utilize vanadium, we would need 1620% of 2010 global reserves. Even if using magnesium-based MOFs, which have demonstrated a great capacity to adsorb CO_2, we would need 14% of 2010 global reserves, which is a considerable amount. Also, extensive mining would be necessary, leading to more potential environmental problems.

In a project sponsored by the DOE and operated by UOP LLC in collaboration with faculty from four different universities, MOFs were tested as possible carbon dioxide removal agents in post-combustion flue gas. They were able to separate 90% of the CO_2 from the flue gas stream using a vacuum pressure swing process. Through extensive investigation, researchers found out that the best MOF to be used was Mg/DOBDC, which has a 21.7 wt% CO_2 loading capacity. Estimations showed that, if a similar system were to be applied to a large scale power plant, the cost of energy would increase by 65%, while a NETL baseline amine based system would cause an increase of 81% (the DOE goal is 35%). Also, each ton of CO_2 avoided would cost $57, while for the amine system this cost is estimated to be $72. The project ended in 2010,estimating that the total capital required to implement such a project in a 580 MW power plant was 354 million dollars.

Other Methods

Many other methods and materials have been discussed for scrubbing carbon dioxide.

- Adsorption
- Regenerative carbon dioxide removal system (RCRS)
- Photosynthesis: e.g. Algae based carbon sink
- Polymer membrane gas separators
- Reversing heat exchangers

Spray Tower

Spray towers are very simple, low-energy wet scrubbers. In these scrubbers, the particulate-laden gas stream is introduced into a chamber where it comes into contact with liquid droplets generated by spray nozzles. These scrubbers are also known as pre-formed spray scrubbers, since the liquid is formed into droplets prior to contact with the gas stream. The size of the droplets generated by the spray nozzles is controlled to maximize liquid- particle contact and, consequently, scrubber collection efficiency. The common types of spray chambers are spray towers and cyclonic chambers. Spray towers are cylindrical or rectangular chambers that can be installed vertically or horizontally. In vertical spray towers, the gas stream flows up through the chamber and encounters several sets of spray nozzles producing liquid droplets. A de- mister at the top of the spray tower removes liquid droplets and wetted PM from the exiting gas stream. Scrubbing liquid and wetted PM also drain from the bottom of the tower in the form of slurry. Horizontal spray chambers operate in the same manner, except for the fact that the gas flows horizontally through the device. A typical spray tower is shown in Figure.

A cyclonic spray chamber is similar to a spray tower with one major difference. The gas stream is introduced to produce cyclonic motion inside the chamber. This motion contributes to higher gas velocities, more effective particle and droplet separation, and higher collection efficiency. Tangential inlet or turning vanes are common means of inducing cyclonic motion.

The scrubbing liquid, usually water for particulate matter removal, is sprayed into the chamber from a series of nozzles located at the top chamber while the air- particle mixture enters the bottom of chamber and flows upward, encountering the droplets formed from the sprays which fall to the bottom by gravity. The droplets remove the particles by scrubbing action, the resulting slurry so formed is collected at the bottom and sent for treatment for removal of collected particles and treated water is recirculated.

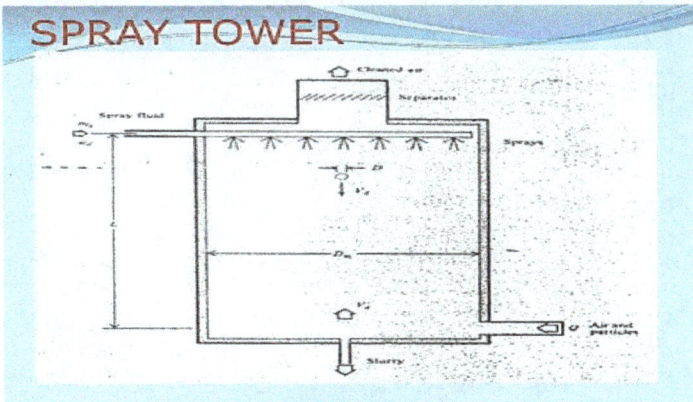

Typical Spray Tower.

Design of Spray Tower

The design of Spray Tower is generally oriented towards collection efficiency of the system which inter-alia include designed upward velocity of air in the spray chamber having regard to Reynolds number, drop formation rates, shape of tower, time for impaction and diffusion along with other parameters, the details of which are reflected as under:

Collection efficiency of spray tower, using following:

- L= length of tower

- D_{sc} = diameter of tower

- V_a = velocity of upward air

- V_d = velocity of dropping droplets

- V_∞ = relative velocity between drops and air

- Re_d = Reynolds number = $10 < Re_d > 700$

- D = diameter of droplet

- ρ_d = density of droplet

- ρ = density of air

- μ = viscosity of air

- n = no of droplets encountered by a group of particles

- N_d = rate of drop formation in number per second

- A_{sc} = area of tower

  Upward velocity of air in chamber:

- $V_a = Q/A_{sc}$, which must not exceed drop velocity V_d to prevent air from carrying drops out of the top of chamber

- F = Force acting on the drop

- $F = \pi * \rho_d * D^3 * g/6 = 5.135 * \rho_d * D^3$

- If spray liquid is water,

- $F = 5135\ D^3$

  If Reynolds number for the drop motion is between 10 and 700:

- $V_\infty = (4.8/\rho*D)*\sqrt{((447*\mu^2)+(\rho*\rho_d*D^3*g/6))} - 20.4*\mu$

  For standard air as gas and water as spray liquid, the above equation becomes:

- $V_\infty = (178.3/D)*\sqrt{(0.7814*10^{-10}+D^3)} - 1.520*10^{-3}/D$

The Reynolds number becomes:

- $Re_d = V_\infty*D/v = 11.50*106*\sqrt{(D^3+0.7814*10^{-10})}-98.06$

If Reynolds number is greater than 700,

$V_\infty = (2.4/D)*\sqrt{(\pi* \rho_d*D^3*g/6*\rho)} = 5.44*\sqrt{(\rho_d*D/\rho)}$

For standard air and water

- $V_\infty = 158 \sqrt{D}$ $Re_d > 700$

Reynolds number is given by

- $Red = 1.020*10^7*D^{3/2}$

Since the drop falls with velocity Vd where

- $V_d = V_\infty - V_a$ (particle travel upward with velocity V_a)

- Time period for impaction, diffusion becomes :

- $L/V_a + L/V_d$, and n becomes:

- $n = (N_d* \pi* D^2/4* A_{sc})*(L/V_a + L/V_d)$ Total drop formation rate is related to mass flow rate of spray fluid as:

- $N_d = 6* ms/\pi* d^3* \rho_d$ n becomes

- $N = (1.5*m_s*L \rho_d*D)*(1/Q+1/(A_{sc}*(V_\infty-V_a)))$

If spray chamber is circular, then

- $A_{sc} = \pi D^2_{sc}/4$

Collection efficiency for a single droplet = ηd, is defined as ratio of no. of particles collected to no. of particles initially contained in the volume swept through by the droplet.To predict the behavior of the particles as they flow around and into the droplet, a particle of given diameter and density will strike the droplet if it lies initially within a certain distance y1 of the axis of motion of the droplet. If it lies away from axis than this, it will pass by the droplet and not collected.

The collection efficiency of the individual droplet due to interception and inertial impaction combined ηdi can be defined as the ratio of the area of circle having radius y1 to the projected area of the droplet. This ratio is modified by an attachment coefficient σ.

From above, efficiency is defined as:

- $\eta di = (\pi * y_1^2 * \sigma)/(\pi * D^2/4) = 4*\sigma*2y_1^2/D^2,$

- $\beta = (5/72)*(\rho_P * d^2 * V_t * C/\mu * D)$

Now considering boundary layer conditions,

- $\eta = 8.811 * \sigma * \sqrt{(v/V_\infty * D)} * ((y_2/\partial_2)^2 - 1/6*(y_2/\partial_2)^4)$

- when $y_2 < \partial_2$

and

- $\eta_{di} = 7.342^* \sigma^* \sqrt{v} / V_\infty^* D$, when $y_2 = \partial_2$

and

- $\eta_{di} = 7.342 * \sigma * \sqrt{v/V_\infty} * D + (2*\sigma)*(y_2/D)$

    $\partial_2/D)^* (3 + 6^* y_2/D + 4^* (y_2/D)^2)$, when $y_2 > \partial_2$

- η = combined efficiency of all the droplet

- $1 - (1 - \eta_{di})^n$

Power requirement

- $W = 9.807 \dfrac{(\rho dx Qsx Ascx L)}{Asc - Q/V\infty} + Q\Delta Ps$

where ΔPs = Pressure drop of water droplets (N/m^2)

Design of Spray Tower for Vsk based Cement Plant

An effort has been made in the present paper by the Author to design a Spray Tower for a 100 TPD (Tonne per day) capacity cement plant based on Vertical Shaft Kiln Technology (VSK) located in Rajasthan.

The monitoring was carried out in respect of relevant designed parameters under all the operating conditions on a time scale and the representative observed values are given here under:

Parameters Observed

- Volume of gases = 25,000 m³/hour=6.94 m³/s

- Temperatures of gases = 100°C

- Inlet SPM concentration = 2000 mg/Nm³

- Density of particle = 1500 kg/m³

Parameters Assumed

- Length of tower = 5 m

- Diameter of tower = 2 m

    $\rho_D = Density\ of\ water\ droplet$

- D= Water droplet = 1 kg/m^3

- D = water droplet size = 2 mm

- msc = flow of water = 0.01 m^3/s

- C= Cunnigatum correction factor =1.0

- σ = Attachment coefficient =1.0

- Δps = Initial pressure drop = 200 N/m^2

- V= viscosity of fluid = 1.55×10-5 m^2/s

Spray Tower Design

- Asc = $\prod$Dc^2 /4

    = 3.14 × 2^2/4

    = 3.14 m^2

- Va = Q/Asc

    = 6.94/3.14

    =2.21 m/s

- Red = 1.020 × 10^7(D)$^{3/2}$

    = 1.020 × 10^7 (0.002)$^{3/2}$

    = 912

- V$_\infty$ = Red × V/D

    $= \dfrac{912\ x1.55\ \times10^{-5}}{0.002}$

    V$_\infty$ =7.02

- V$_d$ = V$_\infty$ − V$_a$

    = 7.06 − 2.21

    = 4.85 m/s

- $n = \dfrac{1.5 \times m_s L[1/Q + 1/Asc(V\infty - Vd)]}{\rho \, \mathrm{d} \times D}$

- $= 1.5 \times 0.01 \times 5[1/6.94 + 1/3.14(7.06 - 4.85)]1 \times 0.002$

- $= 10.8$

- $\eta = 7.342 \times \sigma \, \sqrt{(v/v\infty \times D)}$

- $= 7.342 \times 1\sqrt{(1.55 \times 10^{-5}/7.06 \times 0.002)}$

- $= 0.243$

$$\eta = 1 - (1 - \eta di)^n$$

$$= 1 - (1 - 0.243)^{10.8}$$

$$= 0.95\%$$

Inlet Concentration = 2000 mg/Nm³

Efficiency of spray tower estimated = 95%

Outlet concentration = 0.05 × 2000

$$= 100 mg/Nm^3$$

Which is much below the standard of 250 mg/m³

Hence safe

Now Power requirement = W

$$W = 9.807 \times \frac{\rho \times Qs \times Ascx2}{Asc - Q/V\infty} + Q \times \Delta Ps$$

$$= 9.087 \times 1500 \times 0.01 \times 3.14 \times 5 + 6.69 \times 200$$

$$3.14 - 6.94/7.06$$

$$= 2458 \, \text{watts}$$

References

- Petzow, G. N.; Aldinger, F.; Jönsson, S.; Welge, P.; Van Kampen, V.; Mensing, T.; Brüning, T. (2005). "Beryllium and Beryllium Compounds". Ullmann's Encyclopedia of Industrial Chemistry. doi:10.1002/14356007.a04_011.pub2. ISBN 3527306730

- Gary T. Rochelle (2009). "Amine Scrubbing for CO_2 Capture". Science. 325 (5948): 1652. Bibcode:2009Sci...325.1652R. doi:10.1126/science.1176731

- J.R. Jaunsen (1989). "The Behavior and Capabilities of Lithium Hydroxide Carbon Dioxide

Scrubbers in a Deep Sea Environment". US Naval Academy Technical Report. USNA-TSPR-157. Retrieved 2008-06-17

- Smit, Berend; Reimer, Jeffrey R.; Oldenburg, Curtis M.; Bourg, Ian C. (2014). Introduction to Carbon Capture and Sequestration. Imperial College Press. ISBN 978-1-78326-327-1

- Jones, William P. (1949-11-05). "Development of the Venturi Scrubber". Industrial & Engineering Chemistry. 41 (11): 2424–2427. doi:10.1021/ie50479a020. ISSN 0019-7866

- Griggs, Mary Beth (20 July 2015). "Mercury Scrubbers On Power Plants Clean Up Other Pollutants, Too". Popular Science. Retrieved 30 July 2015

Permissions

All chapters in this book are published with permission under the Creative Commons Attribution Share Alike License or equivalent. Every chapter published in this book has been scrutinized by our experts. Their significance has been extensively debated. The topics covered herein carry significant information for a comprehensive understanding. They may even be implemented as practical applications or may be referred to as a beginning point for further studies.

We would like to thank the editorial team for lending their expertise to make the book truly unique. They have played a crucial role in the development of this book. Without their invaluable contributions this book wouldn't have been possible. They have made vital efforts to compile up to date information on the varied aspects of this subject to make this book a valuable addition to the collection of many professionals and students.

This book was conceptualized with the vision of imparting up-to-date and integrated information in this field. To ensure the same, a matchless editorial board was set up. Every individual on the board went through rigorous rounds of assessment to prove their worth. After which they invested a large part of their time researching and compiling the most relevant data for our readers.

The editorial board has been involved in producing this book since its inception. They have spent rigorous hours researching and exploring the diverse topics which have resulted in the successful publishing of this book. They have passed on their knowledge of decades through this book. To expedite this challenging task, the publisher supported the team at every step. A small team of assistant editors was also appointed to further simplify the editing procedure and attain best results for the readers.

Apart from the editorial board, the designing team has also invested a significant amount of their time in understanding the subject and creating the most relevant covers. They scrutinized every image to scout for the most suitable representation of the subject and create an appropriate cover for the book.

The publishing team has been an ardent support to the editorial, designing and production team. Their endless efforts to recruit the best for this project, has resulted in the accomplishment of this book. They are a veteran in the field of academics and their pool of knowledge is as vast as their experience in printing. Their expertise and guidance has proved useful at every step. Their uncompromising quality standards have made this book an exceptional effort. Their encouragement from time to time has been an inspiration for everyone.

The publisher and the editorial board hope that this book will prove to be a valuable piece of knowledge for students, practitioners and scholars across the globe.

Index